carbon-neutral
architectural design

carbon-neutral
architectural design

Pablo La Roche

CRC Press is an imprint of the
Taylor & Francis Group, an **informa** business

CRC Press
Taylor & Francis Group
6000 Broken Sound Parkway NW, Suite 300
Boca Raton, FL 33487-2742

© 2012 by Taylor & Francis Group, LLC
CRC Press is an imprint of Taylor & Francis Group, an Informa business

No claim to original U.S. Government works

Printed in the United States of America on acid-free paper
Version Date: 20111109

International Standard Book Number: 978-1-4398-4512-7 (Hardback)

This book contains information obtained from authentic and highly regarded sources. Reasonable efforts have been made to publish reliable data and information, but the author and publisher cannot assume responsibility for the validity of all materials or the consequences of their use. The authors and publishers have attempted to trace the copyright holders of all material reproduced in this publication and apologize to copyright holders if permission to publish in this form has not been obtained. If any copyright material has not been acknowledged please write and let us know so we may rectify in any future reprint.

Except as permitted under U.S. Copyright Law, no part of this book may be reprinted, reproduced, transmitted, or utilized in any form by any electronic, mechanical, or other means, now known or hereafter invented, including photocopying, microfilming, and recording, or in any information storage or retrieval system, without written permission from the publishers.

For permission to photocopy or use material electronically from this work, please access www.copyright.com (http://www.copyright.com/) or contact the Copyright Clearance Center, Inc. (CCC), 222 Rosewood Drive, Danvers, MA 01923, 978-750-8400. CCC is a not-for-profit organization that provides licenses and registration for a variety of users. For organizations that have been granted a photocopy license by the CCC, a separate system of payment has been arranged.

Trademark Notice: Product or corporate names may be trademarks or registered trademarks, and are used only for identification and explanation without intent to infringe.

Library of Congress Cataloging-in-Publication Data

La Roche López, Pablo M. (Pablo Miquel)
   Carbon-neutral architectural design / Pablo La Roche.
      p. cm.
   Includes bibliographical references and index.
   ISBN 978-1-4398-4512-7 (hardcover : alk. paper)
   1. Sustainable architecture. 2. Architectural design. 3. Carbon dioxide mitigation. I. Title.

NA2542.36.L3 2012
720'.472--dc23
                                             2011038969

Visit the Taylor & Francis Web site at
http://www.taylorandfrancis.com

and the CRC Press Web site at
http://www.crcpress.com

Contents

Preface xi
Acknowledgments xv
Introduction xiii

Chapter 1   Buildings and Greenhouse Gas Emissions 1

      1.1   Buildings and Greenhouse Gas Emissions 1
      1.2   Anthropogenic Emissions and Climate Change 1
      1.3   Effects of Anthropogenic Emissions on Climate Change 3
      1.4   Greenhouse Gas Emissions and Buildings 7
            1.4.1   Operation Emissions (O_e) 9
            1.4.2   Construction Emissions (C_e) 9
            1.4.3   Emissions from Water (W_e) 10
            1.4.4   Emissions from Waste (W_a) 10
      1.5   Carbon-Counting Tools 10
      1.6   Comparison of Carbon-Counting Tools 11
            1.6.1   Selection of Carbon-Counting Tools 11
            1.6.2   Comparing Emissions from Natural Gas and
                      Electricity 12
            1.6.3   Recommended Carbon-Counting Tools 14

Chapter 2   Carbon-Neutral Architectural Design 17

      2.1   Architectural Design Process 17
            2.1.1   Traditional Architectural Design Process 17
            2.1.2   Computer-Aided Architectural Design Process 18
      2.2   Carbon-Neutral Architectural Design Process 18
            2.2.1   Operation and Energy 23
            2.2.2   Construction 23
            2.2.3   Water 24
            2.2.4   Waste 24
      2.3   Implementation of the Carbon-Neutral Design Process in
            Academia 25
            2.3.1   Carbon-Neutral Architectural Design Process in
                      Beginning-Year Studios 25
            2.3.2   Carbon-Neutral Architectural Design in
                      Advanced Studios 28
      2.4   Integration 60
      2.5   Carbon-Neutral Architectural Design Process in Practice 61

Chapter 3 Thermal Comfort 75

    3.1   Psychrometrics 75
    3.2   Thermal Comfort... 78
            3.2.1   Heat Balance 79
            3.2.2   Variables That Affect Thermal Comfort................... 82
    3.3   Environmental and Comfort Indices 86
    3.4   Comfort Models.. 89
            3.4.1   Physiological Comfort Model 89
            3.4.2   Adaptive Comfort Model 90
    3.5   The Perception of Comfort... 95

Chapter 4 Climate and Architecture 97

    4.1   Climate 97
    4.2   Climate and Architecture 97
    4.3   Climate Zones 100
    4.4   Climate Zones and Energy Codes.. 103
    4.5   Climate Analysis 106
            4.5.1   Building Bioclimatic Chart 106
            4.5.2   Givoni's Building Bioclimatic Chart...................... 108
            4.5.3   Digital Climate Analysis Tools 111
            4.5.4   The Comfort Triangles Chart.................................. 114
    4.6   Vernacular Architecture 114
            4.6.1   Vernacular Architecture in Warm, Humid
                       Climates.. 115
            4.6.2   Vernacular Architecture in Warm and Dry
                       Climates.. 119
            4.6.3   Vernacular Architecture in Temperate Climates 124
            4.6.4   Vernacular Architecture in the Cold Climates........ 128
    4.7   Effects of Climate on Emissions 129
            4.7.1   Assumptions for Simulations 130
            4.7.2   Operation 130
            4.7.3   Construction 133
            4.7.4   Waste 133
            4.7.5   Water 133
            4.7.6   Transportation 134
            4.7.7   Carbon Emissions in the Four Climates 134

Chapter 5 Solar Geometry 137

    5.1   The Sun in the Sky Vault... 137
            5.1.1   Solar Declination and Hour Angle.......................... 137
            5.1.2   Solar Azimuth and Altitude 139

Contents   vii

5.2	Solar Charts	140
	5.2.1 Vertical Sun Path Diagram	141
	5.2.2 Horizontal Sun Path Diagram	142
5.3	Shading the Building	144
5.4	Design of the Shading System	146
	5.4.1 Horizontal Shadow Angle	146
	5.4.2 Vertical Shadow Angle	146
	5.4.3 Shadow Angle Protractor	149
	5.4.4 Example Design Process to Shade a South-Facing Window	150
	5.4.5 Example Design of a Shading for a Southeast-Facing Window	155
5.5	Sundials	157
5.6	Site Analysis	158
5.7	Calculating the Impact of Radiation on Surfaces	161
5.8	Orientation of Buildings	162

Chapter 6   Heat Exchange through the Building Envelope 165

6.1	Heat Transfer through the Building Envelope	165
6.2	Heat Transfer in Buildings	171
	6.2.1 Sensible Heat	171
	6.2.2 Latent Heat	171
	6.2.3 Radiant Heat	174
6.3	Heat Transfer by Conduction	174
	6.3.1 Conductivity	175
	6.3.2 Conductance	175
	6.3.3 Resistance	176
	6.3.4 Thermal Transmittance (U-Value)	177
	6.3.5 Heat Capacity and Specific Heat Capacity	178
	6.3.6 Time Lag	179
	6.3.7 Decrement Factor	179
	6.3.8 Heat Flow by Conduction	180
	6.3.8.1 Use of Insulating Material in Walls, Ceilings, and Floors	184
	6.3.8.2 Use of Air Spaces in Walls, Ceilings, and Floors	185
	6.3.8.3 Increase the Outer Surface Resistance of Walls and Roofs	185
	6.3.8.4 Increase Thermal Resistance of Windows	185
	6.3.8.5 Use of the Thickness of the Architectural Elements as a Regulator of the Building's Indoor Temperature	186

		6.3.8.6	Energy Storage Capacity of Materials...... 187

- 6.3.8.6 Energy Storage Capacity of Materials 187
- 6.3.8.7 Reduce the Temperature Swing Using Materials with High Density and Thermal Capacity 187
- 6.3.8.8 Reducing the Surface Area of the Building 188

6.4 Heat Transfer by Radiation 190
    6.4.1 Concepts 190
    6.4.2 Factors That Affect Solar Radiation 191
    6.4.3 Effects of Solar Radiation 194
    6.4.4 Opaque Components 195
        6.4.4.1 Use of Shading Devices 197
        6.4.4.2 Types of Solar Protection 198
        6.4.4.3 Building Volume 200
        6.4.4.4 Opaque Surface Finish 201
        6.4.4.5 Selection of Absorptive, Reflective, and Emissive Materials for Exterior Surfaces ... 201
        6.4.4.6 Building Components That Are Transparent to Solar Radiation 202
        6.4.4.7 Appropriate Window Selection to Control Solar Radiation 202
        6.4.4.8 Orientation of Buildings and Openings 207
        6.4.4.9 Glazing-to-Surface Ratio 209
        6.4.4.10 Use of Shading Devices 209

6.5 Heat Transfer by Convection 210
    6.5.1 Definition 210
    6.5.2 Air Movement and Infiltration 211
    6.5.3 Controlling the Exchange of Air 213
        6.5.3.1 Seal the Building When $T_e > T_i$ 214
        6.5.3.2 Open the Building When Outdoor Temperature Is Lower than Indoor Temperature ($T_o < T_i$) 214

Chapter 7 Passive Cooling Systems 221

7.1 Definition of a Passive Cooling System 221
7.2 Classification of Passive Cooling Systems 221
7.3 Ambient Air as a Heat Sink (Sensible Component) 223
    7.3.1 Comfort Ventilation 224
    7.3.2 Nocturnal Ventilative Cooling 226
    7.3.3 Smart Ventilation 231
    7.3.4 Effect of Shading on Smart Ventilation 235
    7.3.5 Alternative Methods to Night Ventilate: Green Cooling 236

	7.4	Ambient Air as a Heat Sink (Latent Component: Evaporative Cooling)	242
		7.4.1 Direct Evaporative Cooling	243
		7.4.2 Indirect Evaporative Cooling	246
		7.4.2.1 Givoni–La Roche Roof Pond at UCLA	250
		7.4.2.2 Roof Ponds in a Hot and Humid Climate	253
		7.4.2.3 Cal Poly Pomona Smart Roof Pond with Floating Insulation	254
		7.4.2.4 Cal Poly Pomona Modular Roof Pond	255
		7.4.2.5 University of Nevada, Las Vegas Roof Pond	256
	7.5	The Upper Atmosphere as a Heat Sink: Radiant Cooling	258
		7.5.1 Principles of Radiant Cooling System	258
		7.5.2 UCLA Radiant Cooling System	260
		7.5.3 Zomeworks Double-Play System	263
	7.6	The Earth as a Heat Sink: Earth Coupling	263
		7.6.1 Ground Cooling of the Building by Direct Contact	264
		7.6.2 Ground Cooling of the Building by Earth-to-Air Heat Exchangers	265
		7.6.3 Cooling the Earth	267
	7.7	Applicability of Passive Cooling Systems	267

Chapter 8 Passive Heating 269

	8.1	Applicability of Passive Heating	269
	8.2	Control of Heat Loss	269
	8.3	Passive Solar Heating	270
	8.4	Types of Passive Heating Systems	273
		8.4.1 Direct Gain Systems	273
		8.4.2 Indirect Gain Systems	275
	8.5	Effects of Design Strategies on Emissions	284

References 289

Index 301

PREFACE

Energy used for operation of the buildings is one of the most significant sources of Greenhouse Gas (GHG) Emissions, because of this it is important to understand how to design an envelope that controls energy flows between the interior and the exterior providing natural-passive heating, cooling and lighting.

It is impossible to achieve necessary reductions GHG emissions without implementing significant reductions in the building sector. Even though this can be done using current technologies, it requires a new set of skills and a transformation in how we think about buildings and how we design and produce them. This book provides an introduction to the knowledge that will help to implement appropriate architectural design strategies to reduce emissions and ultimately achieve carbon-neutral buildings, an important part of any coherent set of strategies to reduce anthropogenic impact on climate change. I propose and explain a method to reduce building emissions, with an emphasis on energy and the development of an envelope that works seamlessly with the HVAC system, providing carbon-free heating, cooling and daylighting to its spaces whenever possible. This method is not static, it is dynamic and can be adapted or transformed as needed. It can be used as a guide or starting point that can be further developed and combined in many different ways, including different strategies to achieve carbon neutrality.

Reducing building emissions has become a priority for many people and organizations around the world. In the United States some organizations that have taken important steps to promote the reduction of GHG emissions from buildings are the 2030 Challenge led by Ed Mazria which proposes all new construction to be carbon neutral by 2030, the American Institute of Architecture AIA's 2030 commitment, the American Solar Energy Society and the American Society of Heating, Refrigerating and Air-Conditioning Engineers, ASHRAE. My hope is that this book will help achieve the goal we all seek to reduce building GHG emissions.

ACKNOWLEDGMENTS

I am grateful to the Society of Building Science Educators' Carbon Neutral Design Project and especially to Jim Wasley and Terri M Boake for providing me an opportunity to help me to organize my ideas in this area.

I am also grateful to Energy Design Resources, funded by California utility customers and administered by Pacific Gas and Electric Company, Sacramento Municipal Utility District, San Diego Gas & Electric, Southern California Edison, and Southern California Gas under the auspices of the California Public Utilities Commission, and especially Diane MacLaine and Judie Porter for believing in me and supporting me in the initial phases of the book.

I am also grateful to Maria Martinez-La Roche for her invaluable dedication and support in the development of the general style of the illustrations, diagrams and charts and in the production of many of these, to Eric Carbonnier for developing some of the initial drawings that also contributed to the general style of the illustrations, and to Jillian Schroetinger, Glori Passi, Leslie Cervantes, Nancy Park and Steve Keys for the production of many of the illustrations in the book. Also to Charles Campanella for his dedication and help with many of the energy simulations.

I would also like to thank the many students at the Department of Architecture and the Lyle Center for Regenerative Studies at Cal Poly Pomona University that have in some way been directly or indirectly involved with this project. They always inspire me and motivate me to work harder. I would also like to thank Kyle Brown, director of the Lyle Center and the rest of the staff for supporting much of my recent research, some of which is included in this book.

Finally, I would like to thank John Reynolds, Murray Milne and Marc Schiler for taking some time from their very busy schedules to read my manuscript and provide me with many useful suggestions.

INTRODUCTION

This book is divided into eight chapters. Chapter 1 provides a very brief introduction to the issues of climate change and buildings. Different sources of building emissions are discussed, and how these are produced as a direct result of interactions between buildings and the environment around them. These interactions can be affected by the building fabric and materials, and building inputs and outputs (energy, construction, water, waste). The chapter ends with a comparison of some free carbon counting tools that can be used for residential building design.

A carbon neutral architectural design process (CNDP) is proposed in chapter two and several diagrams were developed to illustrate this process, which has been implemented in architectural education and practice. The basic diagram, which can be adjusted for different types of projects, includes the types of emissions previously discussed: operation, construction, water and waste.

Carbon neutral buildings must achieve thermal comfort with a minimum amount of GHG emissions. Psychrometrics, thermal comfort, the variables that affect it and the different comfort models are discussed in Chapter 3.

Understanding the climate where the building is located helps to identify design strategies that are best adapted to a building's location. Solar geometry is especially important because of the sun's potential to provide heating to the building or its ability to block solar radiation when the building can overheat. Climate analysis is discussed in Chapter 4 and solar geometry is discussed in Chapter 5.

It is important to understand how energy moves through the building envelope. Chapter 6 uses the heat balance equation as a tool to explain how the building fabric can be used to control energy flows by conduction, radiation and convection, which would help to reduce summer overheating and winter underheating. Designing a good envelope that controls heat gains or losses as required is a necessary precondition to better implement passive heating and cooling strategies or mechanical heating and cooling systems.

A passive solar system utilizes the building fabric and materials to heat or cool the building. Passive cooling systems are capable of transferring heat from a building to various heat sinks and they are classified according to the heat sinks that are used to transfer the energy. For each of these, the climates in which they are most effective are indicated in the psychrometric chart. Passive cooling systems are discussed in Chapter 7. Passive heating systems capture solar energy, and then store and distribute it inside the space as needed. Different types of passive heating systems are discussed and even though the principles of passive heating and cooling systems are not new, they are not implemented as much as they should, partly due to lack of knowledge and partly because most energy codes don't consider their favorable effects. Passive heating systems are discussed in Chapter 8.

Much of the information in this book has been developed in my lecture courses in environmental controls at Cal Poly Pomona University and the University of Southern California and my Architectural Design Studios at Cal Poly Pomona. I am a strong believer in the integration of design and lecture courses, so each year as I teach these courses I try to provide more opportunities for my students to implement the concepts that they are learning in the lecture courses in design projects that are integrated in the course and the students in the studios implement more analysis to inform their design decisions. I have also tried to emphasize teaching architectural design strategies over mechanical systems. Even though it is important that students understand how mechanical systems work, I think it is even more important that they learn how to make their buildings work as the mechanical systems, as the knowledge to do this will reduce energy required for heating and cooling, thus reducing building emissions.

1 Buildings and Greenhouse Gas Emissions

1.1 BUILDINGS AND GREENHOUSE GAS EMISSIONS

The greenhouse effect occurs when radiation from the sun passes through the atmosphere and hits the surface of the Earth, which then absorbs it, heats up and emits infrared radiation. Greenhouse gases in the atmosphere absorb most of this infrared radiation and then radiate in the infrared range, which is emitted upward and downward, with part of it absorbed by other atmospheric gases further raising their temperature, part of it escaping to space, and another part reflected downward toward the Earth's surface (Figure 1.1). The surface and lower atmosphere are warmed by the part of the energy that is radiated downward, making our life on Earth possible. Without the greenhouse effect, the Earth's temperature would be about 60°F lower than it is now and too cold for life to exist here.

1.2 ANTHROPOGENIC EMISSIONS AND CLIMATE CHANGE

Some greenhouse gases such as carbon dioxide occur naturally and are discharged to the atmosphere through natural processes and human activities. Other greenhouse gases are created and emitted only through human activities. The main greenhouse gases that enter the atmosphere as a product of human activities are carbon dioxide (CO_2), methane (CH_4), nitrous oxide (N_2O), hydroflourocarbon gases, and sulfur hexafluoride.

Most of the carbon dioxide that enters the atmosphere is a result not only of the burning of fossil fuels (oil, natural gas, and coal), solid waste, trees, and wood products, but also of other chemical reactions (e.g., manufacture of cement). Carbon dioxide is removed from the atmosphere, or "sequestered," when it is absorbed by plants as part of the biological carbon cycle. Methane is given off during the production and transport of coal, natural gas, and oil, from livestock and other agricultural practices, and by the decay of organic waste in municipal solid waste landfills. Nitrous oxide is emitted during agricultural and industrial activities as well as during the combustion of fossil fuels and solid waste. Fluorinated gases such as hydrofluorocarbons, perfluorocarbons, and sulfur hexafluoride are synthetic gases that are generated from a variety of industrial processes and sometimes used as substitutes for ozone-depleting substances. Water vapor is not a man-made greenhouse gas (GHG) and is usually not considered a cause of man-made global warming because it does not persist in the atmosphere for more than a few days (EPA, 2010). Even though other gases have a much higher global warming potential, CO_2 is the most prevalent greenhouse gas and is the most commonly mentioned, sometimes inappropriately.

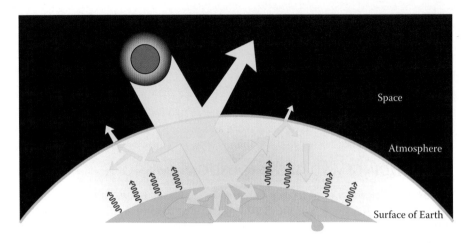

FIGURE 1.1 The greenhouse effect.

TABLE 1.1
GWP of Several Common Anthropogenic Greenhouse Gases

Greenhouse Gas	GWP
Carbon dioxide (CO_2)	1
Methane (CH_4)	21
Nitrous oxide (N_2O)	310
Hydrofluorocarbon gases	6500
Sulfur hexafluoride	23900

The concept of global warming potential (GWP) describes in one number the different capacities to affect climate change. The GWP is defined as the warming influence over a set time period of a gas relative to that of carbon dioxide. The Kyoto protocol assumes a 100-year time period. The GWP of the gases previously described is indicated in Table 1.1.

Another unit used to measure the impact of these gases in one single standard unit is the carbon dioxide equivalent (CO_2e), which also rates these gases based on their 100-year global warming potential, providing a way to compare emissions from different gases using a common unit of measurement (1 kg CO_2e = 1 kg CO_2). The United Nations Environment Programme (UNEP) proposes CO_2e as the common carbon metric for international use in the gathering of data and reporting of the climate performance of existing buildings to (1) support policies under development to reduce GHG emissions from buildings, (2) provide a framework to measure emission reductions in buildings, and (3) establish a system with measurable reportable and verifiable indicators (UNEP SBCI, 2010). CO_2e is the unit of GHG measurement used throughout this book.

1.3 EFFECTS OF ANTHROPOGENIC EMISSIONS ON CLIMATE CHANGE

Life on Earth would not be possible if not for the greenhouse effect; however, an increase in the concentration of greenhouse gases, especially CO_2, originating anthropogenically is affecting climate.

Except for a leveling off between the 1940s and 1970s, the surface temperature of the Earth has steadily increased since 1880, and since the 1990s it has increased at a faster rate in the northern hemisphere than in the southern hemisphere (Figure 1.2). Furthermore, the highest values ever recorded in both hemispheres have been reached in this decade (2000–2010): 2009 was the warmest year ever recorded in the southern hemisphere, and 2010 tied with 2005 as the warmest years ever recorded in the 131 years of instrumental record. However, long-term trends cannot be understood unless several years are compared at a time to reduce "noise." In Figure 1.2 the red line averages 5-year periods at a time, and the warming trend is very clear.

There are many comprehensive reports discussing the origins and effects of climate change; probably the most detailed are those published by the Intergovernmental Panel on Climate Change (IPCC) and available from its website (http://www.ipcc.ch). The IPCC states that "observational evidence from all continents and most oceans shows that many natural systems are being affected by regional climate changes, particularly temperature increases (very high confidence). A global assessment of data since 1970 has demonstrated that it is likely that anthropogenic warming has had a discernible influence on many physical and biological systems" (IPCC, 2007; Parry et al., 2007). Average global surface temperatures have increased by 0.74°C over the past 100 years, and much of the observed rise in sea level (12–22 cm) during the 20th century is probably related to this increase in global mean temperatures. Current data from the National Aeronautics and Space Administration (NASA)

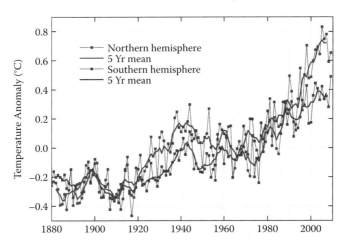

FIGURE 1.2 Hemispheric temperature change. (From NASA Goddard Institute for Space Studies.)

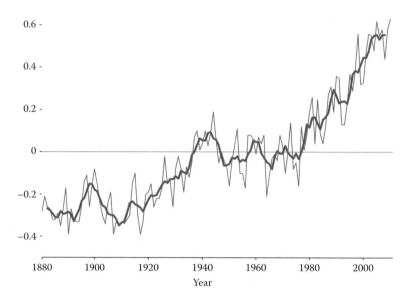

FIGURE 1.3 Global temperature anomalies (°C), 1880–2100. (From NASA Goddard Institute for Space Studies.)

Earth observatory (Figure 1.3) indicate how the 5-year running average continues to increase, especially after 1980. Current IPCC climate models also predict that global temperatures will rise by an additional 1.1 to 6.4°C by the end of the 21st century, depending on the amount of emissions and how the variables interrelate as climate changes. Global mean sea levels are also predicted to rise by 20 to 60 cm by 2100. Sea-level rise threatens coastal communities, while higher temperatures, drought, and flooding will affect people's health and way of life, causing the irreversible loss of many species of plants and animals. There is no doubt that climate change is the most serious environmental threat facing the planet.

Figure 1.4 is a map produced by scientists at the Goddard Institute for Space Studies, which shows the 10-year average (2000–2009) temperature change relative to the 1951–1980 mean. The most significant temperature increases are in the Arctic and the Antarctic Peninsula, and the near-record temperatures of 2009 occurred despite an unseasonably cool December in much of North America. Again, the warming trend is very clear.

In the IPCC report by Yohe et al. (2007), the authors describe the impacts of different climate scenarios as expressed by different global average temperatures on human welfare. In the report they outline the effects on water, ecosystems, food, coastline, and health as well as the time dimension. An important conclusion to be drawn from their report is that even with the lowest scenario of increase in temperature, which is 0.6°C due to past emissions, impacts should be expected. So, even if effective policies are implemented and anthropogenic emissions are quickly reduced to preindustrial levels, our planet will still experience a significant degree of climate change with important worldwide effects on the world's environment, economy, and society. Figure 1.5 illustrates the effects on water, ecosystems, food, coastline, health,

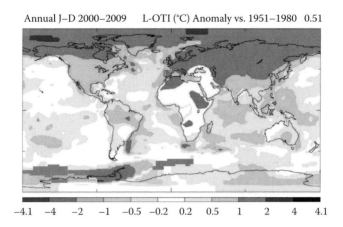

FIGURE 1.4 Ten-year average (2000–2009) temperature change relative to the 1951–1980 mean. (From NASA Goddard Institute for Space Studies.)

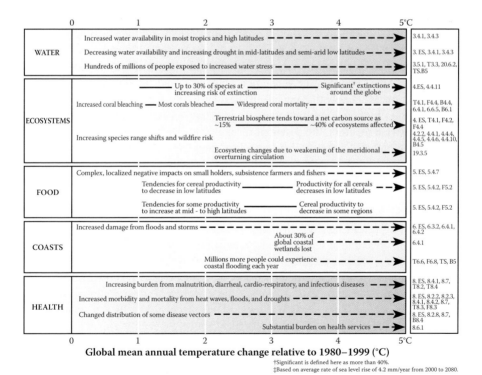

FIGURE 1.5 Effects on water, ecosystems, food, coast, health, and singular events with different scenarios of temperature elevations as measured from the global mean temperature change compared with the 1980–1999 average. (From IPCC 2007.)

and singular events with different scenarios of temperature elevations as measured from the global mean temperature change compared with the 1980–1999 average.

The physical, ecological, social, and economic effects of climate change are numerous and their descriptions are the subject of many specialized publications. According to the IPCC, the extent of climate change effects on individual regions will vary over time and with the ability of different societal and environmental systems to mitigate or adapt to change. Climate change will positively impact some regions and negatively other regions. However, "Taken as a whole, the range of published evidence indicates that the net damage costs of climate change are likely to be significant and to increase over time" (IPCC 2007).

Many of climate change's impacts will affect the built environment. It is virtually certain that over most land areas, warmer and fewer cold days and nights, and warmer and more frequent hot days and nights will produce reduced energy demand for heating and increased demand for cooling. Air quality will decline in cities, and there will be reduced disruptions to transportation due to snow and ice, but there will be effects on winter tourism. It is very likely that warm spells/heat waves will increase in frequency over most land areas and produce a reduction in quality of life for people in warm regions without appropriate housing, negatively impacting the elderly, very young, and the poor. It is very likely that heavy precipitation events will increase in frequency over most areas and produce disruption of settlements, commerce, transport, and societies due to flooding. There will be pressures on urban and rural infrastructures and loss of property. It is likely that the areas affected by droughts will increase, producing water shortages for settlements, industry, and societies, reduced hydropower generation potentials, and the potential for population migration. It is likely that intense tropical cyclone activity will increase producing disruption by flood and high winds, withdrawal of risk coverage in vulnerable areas by private insurers, the potential for population migrations, and loss of property. It is likely that there will be increased incidence of extreme high sea level (excludes tsunamis) that will increase the costs of coastal protection versus the costs of land-use relocation and increasing the potential for movement of populations and infrastructure.

In general, heat-related mortality in many urban centers will increase. Coasts will be exposed to increasing risks, including coastal erosion; millions of people will be flooded every year due to sea-level rise by the 2080s, especially in densely populated, low-lying areas, such as the mega-deltas of Asia and Africa and small islands. The most vulnerable industries, settlements, and societies are usually those in coastal and river flood plains, those whose economics are closely linked with climate-sensitive resources, and those in areas prone to extreme weather events, especially where rapid urbanization is occurring. Poor communities can be especially vulnerable, in particular those concentrated in high-risk areas. They tend to have more limited adaptive capacities, and are more dependent on climate-sensitive resources such as local water and food supplies. We have to prepare ourselves for these changes, and architecture will most certainly have to adapt to climate change. A recent example is New Orleans for which we must now design thinking that part of the house is expendable and will disappear in a flood.

1.4 GREENHOUSE GAS EMISSIONS AND BUILDINGS

It is difficult to determine the exact amount of energy used by the building sector because it is usually not considered an independent sector with its own data. Furthermore, there is a lack of consistent data from the different entities that collect this information. Different sources estimate GHG emissions from the building sector to represent around 33% of emissions worldwide, according to the United Nations Environment Programme (UNEP) Sustainable Buildings and Climate Initiative (SBCI) (UNEP SBCI, 2010). In many industrialized more developed nations, this number is higher, and in the United States, estimates range from 43% (Brown, Stovall, and Hughes, 2007) to 49% (Mazria and Kershner, 2008) of total emissions. All of these are significant, and building emissions are always the single largest source, higher than emissions from industry and transportation. Furthermore, these include emissions from energy use only in the building. If other building-related emissions are considered, the number would be significantly higher. The IPCC's 4th Assessment Report (IPCC AR-4) estimated that building-related GHG emissions reached 8.6 billion metric tons (t) CO_2e in 2004 and could nearly double by 2030, reaching 15.6 billion tCO_2e under its high-growth scenario. This means that just from the sheer quantity of these emissions, to have any real impact on climate change it is crucial that emissions from the building sector be addressed. Furthermore, according to this same report, buildings provide the most economic mitigation potential for reduction of GHG emissions and the largest potential for reducing GHG emissions and are relatively independent of the price of carbon reduction (cost per tCO2e) applied (Figure 1.6). With proven and commercially available technologies, the energy consumption in both new and existing buildings can be cut by an estimated 30–50% without significantly increasing investment costs.

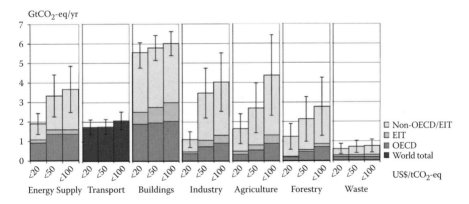

FIGURE 1.6 Estimated economic mitigation potential by sector and region using technologies and practices expected to be available in 2030. The potentials do not include nontechnical options such as lifestyle changes. Organization for Economic Cooperation and Development = OECD and Economies in Transition = EIT. (From Parry, M. L., Canziani, O. F., Palutikof, J. P., van der Linden, P. J., and Hanson, C. E. (Eds.), *Contribution of Working Group II to the Fourth Assessment Report of the Intergovernmental Panel on Climate Change*, Cambridge, UK: Cambridge University Press, 2007, Figure SPM.6. With permission.)

The IPCC states that measures to reduce GHG emissions from buildings fall into one of three categories: (1) reducing energy consumption and embodied energy in buildings; (2) switching to low-carbon fuels, including a higher share of renewable energy; or (3) controlling the emissions of non-CO_2 GHG. A very large number of technologies that are commercially available and tested in practice can substantially reduce energy use while providing the same services and often substantial co-benefits.

In another very comprehensive report published by the American Solar Energy Society (ASES) to investigate different climate change mitigation strategies (Brown et al., 2007), scientists estimated that 43% of U.S. carbon dioxide emissions result from the energy services required by residential, commercial, gas, and industrial buildings and that there was a huge potential for reduction in emissions by increasing the efficiency of buildings: 688 MtC/yr in 2030 compared with 63 for concentrating solar, 63 for photovoltaics, 181 for wind, 53 for biofuels, 75 for biomass, and 83 for geothermal.

It is clear that significant reductions in emissions will be possible only if reductions in emissions from reductions in energy consumption in buildings are implemented. However, buildings emit GHGs also via their operation, construction, water use, and waste creation (Malin, 2008).

Emissions are produced as a direct result of interactions between the building and the external environment that surrounds it. These interactions can be affected by the building fabric and materials (Figure 1.7) or the building's inputs and outputs (operation, construction, water, waste):

$$T_{be} = O_e + C_e + W_e + W_a \tag{1.1}$$

where

T_{be} = total building emissions
O_e = operation emissions (energy)
C_e = construction emissions
W_e = water emissions
W_a = waste emissions

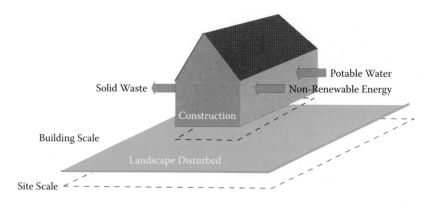

FIGURE 1.7 Building emissions and environmental interactions.

Buildings and Greenhouse Gas Emissions

Other emissions such as those from transportation are affected by building location but are not produced by the building itself or as a result of any of the building's inputs and outputs and are thus not considered part of a building's emission in this book.

1.4.1 Operation Emissions (O_e)

Operation emissions are produced from energy used to operate the buildings. They are the single largest source of building-related GHG emissions, 80% according to UNEP SBCI (2010). UNEP proposes, and most GHG protocols have adopted, a classification of GHG in terms of scopes that measure emissions only during the use phase (e.g., operation, maintenance, and retrofits) of the building. UNEP proposes three scopes that are commonly used in most GHG protocols.

Scope 1: Direct, On-Building-Site or On-Building-Stocks GHG Emissions
Direct on-site emissions result from sources within the boundaries of the building or building stocks under study that can be quantified by the reporting entity, including stationary combustion emissions from fuels and combustion from boilers, furnaces or turbines, and fugitive emissions from intentional or unintentional releases.

Scope 2: Indirect On-Building-Site GHG Emissions
Indirect on-building-site GHG emissions are sometimes called indirect energy emissions. They are a consequence of the activities that occur outside the building site, for example, activities at a community power plant in the process of providing the energy consumed on-building site. Scope 2 emissions include the emissions produced in the process of producing energy that is used in the building. These include all GHG emissions associated with the overall generation of purchased energy such as electricity or steam for cooling, ventilation, or heating.

Scope 3: Other Indirect GHG Emissions
Scope 3 addresses indirect emissions not covered in Scope 2 activities that are relevant to building performance not included in the common carbon metric. Examples of these include upstream and downstream emissions related to the before-use phase of the buildings (e.g., raw material extraction for construction materials), transportation-related activities during all stages of the building life cycle, and after-use activities such as reuse and recycling of building components. Some protocols (DEFRA, 2010) divide transportation into two groups and include owned transportation in Scope 1 and other transport-related activities such as commuting in Scope 3 (Figure 1.8).

1.4.2 Construction Emissions (C_e)

These can be produced during the fabrication of the materials used in the building, during the transportation of materials to the building, and during the construction of the building.

 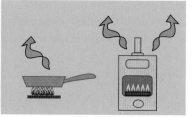

Sources of Direct Emissions   Sources of Indirect Emissions

FIGURE 1.8 Direct and indirect emissions.

1.4.3 Emissions from Water (W_e)

Water consumption in the building also generates GHG emissions when it is pumped from the source and treated for consumption. As the wastewater from the building is treated to remove physical, chemical, and biological contaminants, energy is generated in this process.[*] These emissions are different from domestic water heating emissions, which are included in the energy and operation of the building.

1.4.4 Emissions from Waste (W_a)

Solid waste (not transported in water) coming from the building must be moved from the building and treated, which generates emissions. In addition to this, the waste that arrives at the landfill usually produces methane while it is decomposing.

1.5 CARBON-COUNTING TOOLS

The concept of carbon footprint has been defined in several ways but usually involves an estimate of the carbon dioxide emissions that an individual is directly responsible for over a given period of time (Padgett, Stenemann, Clarke, and Vanderbergh, 2008), and several carbon footprint calculators are available online to determine personal carbon emissions. These calculators can be used during the architectural design process, these types of calculators can help to provide a "feel" for the numeric relationship among lifestyle, consumption patterns, and emissions (Boake, 2010).

Other definitions of carbon footprint take into effect life-cycle analysis. The Carbon Trust describes it as "... a methodology to estimate the total emission of greenhouse gases (GHG) in carbon equivalents from a product across its life cycle from the production of raw material used in its manufacture, to disposal of the finished product (excluding in-use emissions)" (Carbon Trust, 2007).

Some authors (e.g., Wackernagel et al., 2005) see the carbon footprint as part of the ecological footprint (EF). However, they are different because the ecological footprint indicates a demand for resources and is expressed in units of area, hectares, or even planets, while the EF developed by Rees and Wackernagel (Rees,1992) measures how much bioproductive area (whether land or water) a population would

---

[*] In practice these may vary significantly by location and climate.

Buildings and Greenhouse Gas Emissions

require to sustainably produce using prevailing technology all the resources it consumes and to absorb the waste it generates. The carbon footprint is an emission—not a demand—and is measured in units of mass, not units of area.

Not only do carbon footprint calculators generate results in different units (e.g., planets versus kilograms), but further complications also arise from the previously discussed fact that there are several greenhouse gases with different warming potentials. This is why CO_2e is used as the common unit of measurement to normalize results and permit direct comparison between different sources and types of emissions. For example, emissions from water use can be compared with emissions from energy used in the building.

The first step to mitigate greenhouse gas emissions from buildings is to be able to count them. If carbon counting is integrated in the design process, the impact of architectural design strategies can be evaluated more easily to develop buildings with reduced emissions. To determine which of these could be used in the architectural design process, 40 greenhouse gas calculators and energy modeling software programs were compared in the main areas in which buildings are responsible for carbon emissions: operation, water, construction, and waste. Most carbon-counting tools offer the possibility of determining emissions from energy use in buildings or from transportation. Fewer of them can determine emissions due to water, waste, or construction. A handful of these even allow calculation of the carbon impact generated from the food eaten in the building. Because many of these tools accounted for transportation, it was also included in this analysis (La Roche, 2010; La Roche and Campanella, 2009). These tools are compared in the next section.

1.6 COMPARISON OF CARBON-COUNTING TOOLS

To develop this carbon-neutral design process (CNDP), several carbon-counting tools were compared to determine their performance and potential utility in this process. These tools can be used to convert the units obtained in the different areas to a common unit of measure that would permit their comparison (CO_2e).

1.6.1 SELECTION OF CARBON-COUNTING TOOLS

The 40 tools are listed in Figure 1.9. The first column lists the names of the tools, and the rest of the columns provide the emission areas: operation, transportation, waste, construction, and water. The gray rectangles indicate that the tool can be used to calculate emission in that particular area. Most of the tools calculate emissions from some type of operational energy and some mode of transportation (usually automobile or plane), but few of them account for those from waste, construction, or water. The full table with these tools is also posted on the Carbon-Neutral Design Studio Project website (http://www.aia.org/carbonneutraldesignproject), in the tools section. In addition to an image of a representative input and output screen, the table provides (1) the URL address, (2) the areas for which the tool calculates carbon emissions and each tool's subareas, (3) the ease of use on a three-point scale (easy, moderate, difficult), (4) the time needed to complete the information, (5) the units used in the output, and (6) general comments.

1.6.2 Comparing Emissions from Natural Gas and Electricity

To determine their precision, some of these tools were compared using the same input. Because many tools can calculate emissions from electricity and gas (Figure 1.9), several of these were compared by providing them with the same inputs and comparing their outputs. Only tools with the option to accept a similar input and to produce an output in the same units were selected for comparison. An annual consumption of 12,000 kWh was used as the input for electricity and 2,610 m^3 (921.6 Therms) for natural gas.

Energy used by the building can be expressed as site energy or source energy. Source energy is the total amount of energy that is required to operate the building and it includes all transmission, delivery, and production losses, which can be

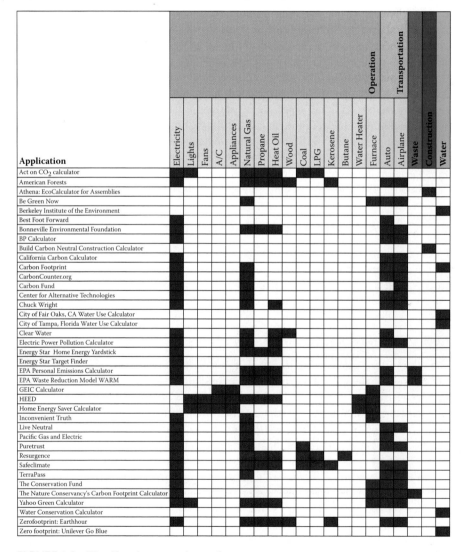

FIGURE 1.9 The 40 carbon-counting tools.

a substantial amount of the total used on site. Site energy is the amount of heat and electricity actually consumed by a building on site and it is the amount measured by the utility company. All of these calculators use site energy for their calculations.

A total of 14 calculators were compared, all of which generated different CO_2e values. For an input of 12,000 kWh, the emissions ranged from 2,673 kg CO_2e /year to 8,182 kg CO_2e /year. The average was 5,286 kg CO_2e /year, or 0.44 kg CO_2e/kWh, ranging from 0.22 to 0.68 kg/kWh (Figure 1.10). As a reference, the average emission factor for grid electricity in the United States is 0.62 kg CO_2e/kWh (eGrid, 2007), and in the United Kingdom it is 0.43 kg CO_2e/kWh (DEFRA, 2005).

There was less variation in the outputs from natural gas. For an input of 2,610 m³ of natural gas, the outputs ranged from 4,182 kg CO_2e/year to 5,465 kg CO_2e/year. The average was 5,043 kg CO^2e/kWh, or 0.187 kg CO_2e /kWh (5.47 kg CO_2e/therm). The emissions range from 4.54 to 6.13 kg CO_2e /therm (Figure 1.11), and the average

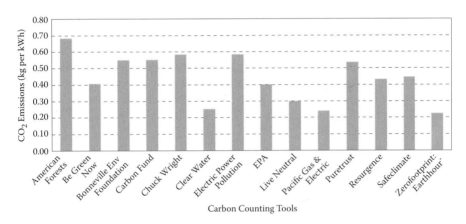

FIGURE 1.10 Conversion factor for electricity normalized in kg CO_2e/kWh for the 14 calculators.

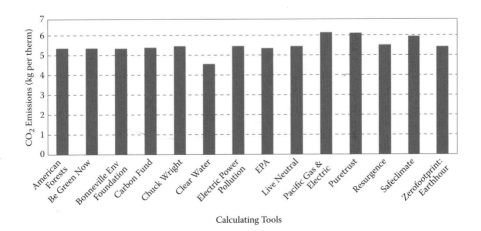

FIGURE 1.11 Conversion factor for natural gas normalized in kg CO_2e/therm of gas.

is relatively close to 5.43 kg CO_2e/therm (0.19 kg CO_2e/kWh), which is the fuel emission factor for natural gas proposed by DEFRA (2005).

Unfortunately, most of these calculators are not transparent, and it is not easy to determine the calculation methods and conversion factors used by their engines, most of which probably implement very simple numerical operations. The variability in the results generated by the calculators indicates that it is better to determine appropriate site-specific conversion factors and to multiply this factor by the calculated or recorded energy used in the building.

1.6.3 Recommended Carbon-Counting Tools

Based on the analysis of the tools, I selected several that can be used to count residential emissions; they are indicated in Table 1.2. When these tools were compared, there was no specific carbon-counting tool specifically designed for buildings. Recently a new tool called green footstep (http://greenfootstep.org/), developed by the Rocky Mountain Institute, incorporates many features that can be performed with all of these tools, but most of the calculations must still be performed outside of the tool itself. Another carbon-counting tool designed specifically for architecture is under development at HMC Architects in California as a smartphone application and website.

TABLE 1.2
Recommended Carbon-Counting Tools for Residential Building Design

Carbon Emissions	Tool
Operational energy	HEED (Home Energy Efficient Design)
	Provides separate results for heating, cooling, lights, appliances, fans, and water heater, enabling the designer to better determine the most effective strategy. Yearly energy results in the building performance screen (BEPS) can be multiplied by a conversion factor to obtain CO_2e or can be obtained directly from the program.
	http://www2.aud.ucla.edu/energy-design-tools/
Construction	Build Carbon Neutral
	Provides a very rough estimate. The total must be divided by estimated life of the building to obtain CO_2e/year.
	http://buildcarbonneutral.org/
	or
	Athena EcoCalculator for Assemblies
	Provides a more detailed analysis but is not available for all regions.
	http://www.athenasmi.org/tools/ecoCalculator/index.html
Water	The amount of energy embedded per unit of volume of water must be multiplied by the CO_2e conversion factors and the yearly building water use to obtain CO_2e.
	A CO_2e factor per million gallons (1,331 lbs of CO_2e/MG (0.002285 kg of CO_2e/l) was used (Southern California Factor). The water use in the building can be determined with many tools, for example, goblue.zerofootprint.net
Waste	EPA Personal Emissions Calculator (see the Waste section in this chapter) for simple analysis permits the designer to model the effects of reducing home waste and recycling.
	http://www.epa.gov/climatechange/emissions/ind_calculator.html
	EPA WARM model for a more detailed analysis:
	http://yosemite.epa.gov/oar/globalwarming.nsf/content/ActionsWasteWARM.html
Transportation	To obtain the emissions from transportation per household, the number of miles traveled per household is multiplied by the CO_2e factor per mile for that mode of transportation (train, bus, automobile). For example, for automobiles, the number of miles traveled is divided by the fuel efficiency of the automobiles and then multiplied by the emissions factor per gallon of fuel. Only automobile was considered with a factor of 2.32 kg of CO_2e per liter and a fuel efficiency of 9.36 km/l (19.56 lbs of CO_2 per gallon of gas and 22 miles per gallon) can be used. The number of miles traveled per year must be estimated and proposed.

2 Carbon-Neutral Architectural Design

To reduce anthropogenic emissions and have any real impact on climate change, it is necessary to reduce emissions from the building sector, and to do this it is necessary to know how to reduce them. This can be achieved through a systematic process: the carbon-neutral design process (CNDP).

Architects should know how to design carbon-neutral buildings with a reduced environmental impact. However, they have not been trained to do this. This training begins in architecture school, and this chapter discusses the CNDP implemented by the author in several architectural design studios, which is beginning to be used in practice. The objective of this CNDP is to help a designer reduce building-related anthropogenic greenhouse gas emissions. Many designers are aware of the impact of building operation on emissions, but emissions can also originate during the fabrication of the building, from the supply of water to the building and from the disposal of waste from the building. Emissions from transportation are also affected by the building's location, to include consideration of neighborhood and local or regional planning issues.

2.1 ARCHITECTURAL DESIGN PROCESS

2.1.1 Traditional Architectural Design Process

The architectural design process is intimately related to the process of problem solving, which in very general terms means that we need something but do not know exactly how to obtain it. The design process is also an iterative problem-solving process in which sequences of subproblems that are not well defined are solved. During the architectural design process the documents to generate a building, which must satisfy many different criteria, are produced. Traditionally this process begins with a series of diagrams and sketches that illustrate the architectural concept. These sketches are progressively refined with the addition of different criteria until the final proposal is developed.

Rittel (1970) proposed the existence of an alternating cycle, which repeats itself continuously in the design process. This process involves two types of mental activities: the generation and the reduction; between these cycles there are periods of work with simple, routine activities in which the generation or reduction of ideas decreases in intensity. In general terms, during the generation stage we propose alternatives, and during the reduction phase we evaluate these alternatives. Thus, the architectural design process is a complicated one in which the architect generates and evaluates multiple solutions to finally produce a sufficiently refined object that fulfills the requirements of sometimes conflicting ideas or programs (Figure 2.1).

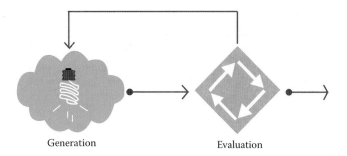

FIGURE 2.1 Alternating generation–evaluation cycle in the design process.

2.1.2 Computer-Aided Architectural Design Process

A computer-aided design (CAD) tool is a program or set of computer tools that substitute for or add to traditional design tools used in the architectural design process. A CAD tool can be any computer program that helps the architect solve an issue during the architectural design process. The ultimate objective of all CAD tools is to improve the product by improving the design process (La Roche, 1995).

The ideal computer design tool should be easy to use but precise so that it will have a larger base of users that will be able to generate a larger amount of buildings in a reduced amount of time. Very simple design (VSD) tools, proposed by La Roche and Liggett (2001), should be very easy and simple to use so that more designers can use them in more projects, which would improve the overall quality of a larger building stock. VSD tools should include at least four of the following six characteristics:

1. **Easy to use.** They must have a graphical user interface and should provide results with a minimum amount of data, usually not more than six data inputs.
2. **Easy to learn.** Use of the tool should be intuitive so that the designer can be up and running quickly with it.
3. **Sufficiently precise.** The results should have sufficient precision to be useful during the initial stages of the architectural design process.
4. **Fast.** The process of inputting data and processing the information to generate results should not slow down the design process.
5. **Accessibility.** The tools should be free or very inexpensive so that they are available to a large number of users. The World Wide Web is especially useful as a sharing tool accessible to millions of users all over the world.
6. **Information exchange with other VSD tools.** The tools should permit the flow of information with other VSD tools to help maintain the continuity of the design process.

2.2 CARBON-NEUTRAL ARCHITECTURAL DESIGN PROCESS

It is important that architects learn how to reduce a building's greenhouse gas (GHG) emissions. Currently, no single tool is available to calculate emissions from the

Carbon-Neutral Architectural Design

building in all areas; however, a combination of methods, tools, and techniques can be used, and this section discusses several of them.

In this process the designer must first generate an emission baseline, which is expressed in kilograms of CO_2e per square meter per year (kg $CO_2e/m^2/yr$). In a carbon-neutral design process, all the outputs from the different sources must be converted to this common measuring unit. The baseline is the desired goal and is affected by the physical and social environments. It includes society's needs, as expressed in a program, and budgetary constraints and code requirements, combined with environmental pressures to maintain healthy physical and biological environments. These pressures will continue during the development of the project, construction, and its operation and will affect them from the initial conception of the idea to its construction, operation, and demolition. During this process the designer continually generates and evaluates ideas, compares them with the goals and the baseline, and adjusts them accordingly (Figure 2.2). The product of the design process is the project, which provides the necessary documents for construction of the building and must perform better than the baseline. After the project is concluded and in operation, a post-occupancy evaluation (POE) will help to determine if the building is operating according to project predictions.

The ideal baseline is zero emissions; however, this is difficult to achieve because it is not possible to eliminate all of a building's emissions. Energy and water will always be used, and some waste will always be produced (refer to Figure 1.7). Consumption must first be reduced by implementing appropriate design strategies, and then carbon sinks should be implemented. Just as there are sinks for energy, there can also be carbon sinks. A building is more or less "dirty" depending on the magnitude of these emissions. A dirtier building emits more GHG than a cleaner building. A regenerative building goes a step further and can absorb more emissions than it produces, becoming a sequesterer of emissions. A sequester, in addition to building emissions, also absorbs the equivalent of the emissions produced during its construction and those produced by water use and generation of waste. A building can be an emitter or a sequester depending on the numerical relationships between the emissions from operation, waste, water, and construction (the latter distributed over the building's lifetime) and the magnitude of the carbon sinks. A dirtier building is located toward

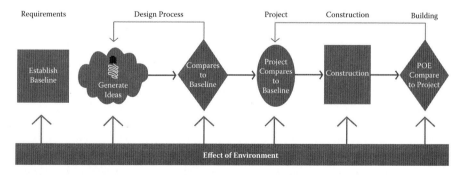

FIGURE 2.2 General phases in the carbon-neutral design project.

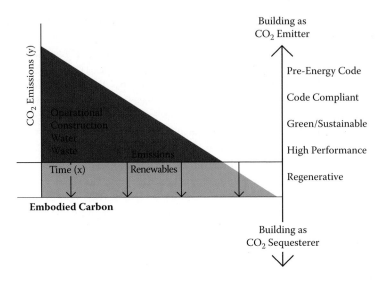

FIGURE 2.3 Classification of building according to GHG emissions.

the top of Figure 2.3, whereas a cleaner building is located toward the bottom of the figure, and a regenerative building is located in the green area.

To better illustrate the CNDP, diagrams that serve as road maps during the design process were developed. Figure 2.4 illustrates a proposal developed by the author and implemented in several carbon-neutral design studios at California State Polytechnic University, Pomona (hereafter referred to as Cal Poly Pomona). This diagram can be continually refined and updated as needed, per specific requirements, thus becoming a road map during the design process.

In this diagram (Figure 2.4), the four horizontal bands are the sources of emissions previously discussed: operation, construction, water, and waste. The first column describes the baseline requirements for the different areas (operation, construction, water, and waste), and the final column describes what was actually achieved for each area, the outcomes. The columns between these two are roughly equivalent to scales in the process, from the regional–urban scale to the site, the building envelope, and finally the interior of the building. Different knowledge areas that discuss emission reduction strategies are located at the intersection of the columns and the bands and are represented with colored rectangles. Some of these knowledge areas can reduce the carbon intensity of the building, while other knowledge areas can actually perform a function in the building that is actually the equivalent of absorbing the emissions, becoming carbon sinks. Knowledge areas that reduce the intensity of the carbon emissions are enclosed in rectangles, while areas that are carbon sinks have a thicker border around these rectangles.

This book is organized following the diagram, indicating in each knowledge area the chapter that discusses it and providing information that can be used to reduce emissions in specific areas (Figure 2.5). These chapters are not meant to be exhaustive references on each topic, but together they provide an overview of the knowledge necessary to design carbon-neutral buildings. In this first edition I am concentrating

Carbon-Neutral Architectural Design

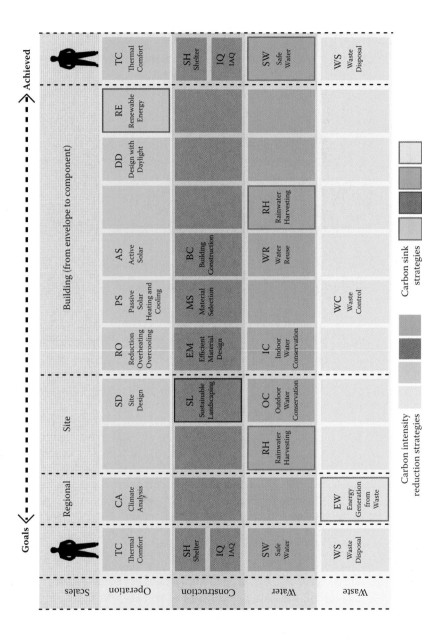

FIGURE 2.4 Carbon-neutral design process.

22 Carbon-Neutral Architectural Design

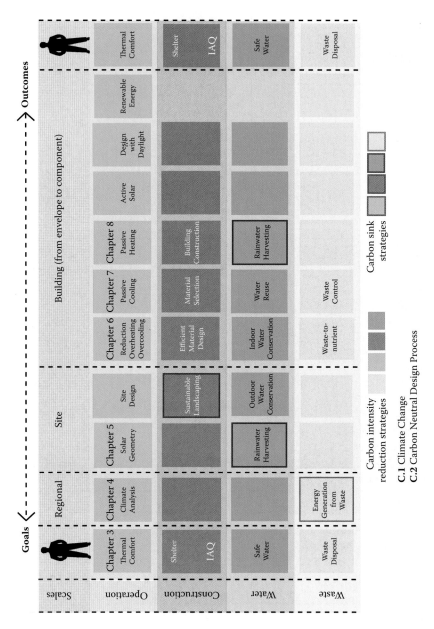

FIGURE 2.5 Chapters of the carbon-neutral design process developed in this book.

Carbon-Neutral Architectural Design

on emissions from operation, which are responsible for at least 60% of all GHG emissions originated over the life of the building. I hope to continue developing these and adding more sections in future editions of the book.

2.2.1 Operation and Energy

Operation emissions are generated during the production of energy necessary to operate the buildings. All of the knowledge areas listed here are included in the green "operation" row and provide knowledge that can contribute to reduce these emissions.

Thermal comfort (TC). Carbon-neutral buildings should provide healthy and thermally comfortable environments with minimum GHG emissions. A carbon-neutral building must achieve a certain level of comfort with a minimum amount of GHG emissions and environmental impact. Thermal comfort is discussed in Chapter 3.

Climate analysis (CA). The analysis of the climate data where the building is located permits to propose design strategies that would help to achieve thermal comfort with minimum energy use in the building. Climate analysis is discussed in Chapter 4, and solar geometry is discussed in Chapter 5.

Site design (SD). The environment directly around the building also affects a building's heat balance and can be designed to reduce emissions from operation, waste, and water.

Reduction of overheating and overcooling (RO). This deals with the use of the building fabric to control energy flows to reduce summer overheating and winter overcooling. This is a necessary first step before implementing passive heating and cooling strategies or mechanical heating and cooling systems. Reduction of overheating and overcooling is discussed in Chapter 6.

Passive solar (PS). A passive solar system uses the building fabric and materials to heat or cool the building. Passive solar systems are discussed in Chapter 7 (cooling) and Chapter 8 (heating).

Active solar (AS). Active solar systems designate mechanical devices that collect solar energy in the form of heat and then store it for later use. The working fluid is usually water or air.

Design with daylight (DD). Daylight reduces use of electrical light and thus the use of electricity and the emissions associated with it. Furthermore, it provides many additional psychological benefits.

Renewable energy (RE). Renewable energy is natural energy that is not limited in supply and will never run out. Using renewable energy can be considered a carbon sink because it offsets the use of nonrenewable energy.

2.2.2 Construction

Emissions from construction can be produced during the fabrication of the materials used in the building, during the transportation of materials to the building, and during construction of the building.

Shelter (SH). One of the basic goals of a building is to provide appropriate shelter for the occupants and their activities.

Indoor Air Quality (IQ). Another fundamental goal is to provide healthy indoor environments.

Sustainable landscaping (SL). Well-designed sustainable landscaping can absorb carbon emissions or reduce GHG emissions from the building. Landscaping can generate carbon emissions by the disturbance of the soil but can also act as a carbon sink (e.g., planting of rapid-growth plants).

Efficient material design (EM). Designing the building efficiently to reduce waste and optimize the use of materials (e.g., using modular design).

Material selection (MS). Selection of low-energy, low-carbon materials is preferable because it is easier to offset when emissions are low.

Building construction (BC). Building construction can be optimized to reduce emissions during construction processes from the flow of materials and people and to minimize waste and use of materials. Reuse of a building or part of a building will significantly reduce emissions from construction.

2.2.3 Water

Water consumption in the building also generates GHG emissions during pumping and treatment from the source. The wastewater from the building must also be treated to remove physical, chemical, and biological contaminants.

Safe water (SW). Provision of safe drinking water is a basic human requirement. In addition to this requirement, excessive water use must be eliminated.

Rainwater harvesting (RH). Rainwater harvesting is the capture and storage of rainwater to provide potable water for drinking or irrigation or to refill aquifers. Rainwater harvesting is considered a carbon sink because it offsets consumption of potable water, which not only is a very limited resource but also generates GHG emissions during its distribution.

2.2.4 Waste

Solid waste (not transported in water) coming from the building must be transported away from the building and treated, generating emissions in this process. In addition to this, organic waste decomposing in the landfill will usually generate methane. If this methane is not harnessed as energy, it will escape to the atmosphere as a potent greenhouse gas.

Energy generation from waste (EW). Methane tapped from a landfill can be used to generate energy. This energy is renewable as long as waste continues to accumulate, and it is a carbon sink because it is using a GHG to generate energy and to reduce its effect in this process.

Composting (CM). Composting is the purposeful aerobic biodegradation of organic matter into mostly organic material that takes the form of a dark, crumbly substance. Through composting, the amount of waste going out of the building is reduced.

Waste control (WC). Waste prevention and recycling reduce greenhouse gases by reducing the amount of new material that must be processed and the amount of material that goes back to the waste stream.

2.3 IMPLEMENTATION OF THE CARBON-NEUTRAL DESIGN PROCESS IN ACADEMIA

Architectural education is a complex process that, in addition to traditional lectures, requires hands-on implementation of lecture material in studios. This means that the knowledge needed to reduce emissions should be embedded in both lecture and studio courses. Introductory design studios provide the initial exposure of students to these concepts, lectures provide opportunities for in-depth learning of issues and concepts, and advanced topic studios provide opportunities to implement concepts learned in initial studios and lecture courses.

2.3.1 Carbon-Neutral Architectural Design Process in Beginning-Year Studios

It is critical that environmental technology be taught at the introductory level, and the design studio is central to this learning process. If students are introduced to sustainability during the beginning years, these concepts can become an integral part of their design repertoire and will probably be implemented in other projects (La Roche, 2008). Undergraduate students in the Department of Architecture at Cal Poly Pomona are introduced to sustainable architectural design in the spring quarter of their second year in a set of required design studio and lecture courses, which are complemented by construction and history courses. In the lecture portion, general sustainability concepts are introduced with special emphasis on the control of natural forces: the sun and the wind. These concepts are incorporated in design projects in the studio portion of the course, and underscore an understanding of energy, the main "architectural factor" that affects climate change. At this introductory level this is achieved by understanding shading, daylighting, and air movement in buildings.

Because a designer has the capacity to simultaneously process only a limited number of pieces of information, it is important to reduce the number of issues that the designer, in this case a student, has to think about simultaneously to a manageable amount, permitting to better resolve the issues. In most of my environmentally responsive studios, including the second-year studio, the importance of some design variables is reduced to give more importance to the sustainable design variables.

Several diagrams were developed over several studios to illustrate the design process, which usually begins with an analysis at the urban scale or neighborhood and continues with the site analysis, the building envelope, and the interior of the building (Figure 2.6). Emphasis is on daylight and ventilation at the building scale but, again, at a very introductory level. Daylight is studied with digital tools or simple

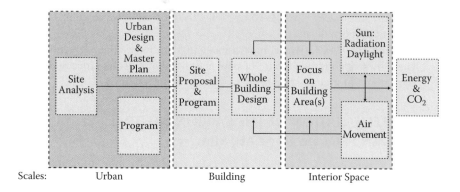

FIGURE 2.6 Second-year studio design process.

physical models and illuminance meters, and ventilation is studied with analog tools, usually with a simple class-made wind tunnel.

These analyses permit students to better understand the physical phenomena that affect thermal performance. There are several reasons for doing analog work before doing digital work: (1) students should already have a sense of some of the physical phenomena before doing the digital work; (2) a more detailed level of calculation, which might not be attainable with the analog tools, is required; (3) some concepts cannot be calculated using analog tools (e.g., carbon emissions). As the student progresses through school, more digital analysis is done substituting some of the analog analysis (Figure 2.7). Furthermore, each year as students become more computer savvy and software more powerful, inexpensive, and easier to use, it will become easier and more common to use digital tools in the beginning design stages. However, the haptic value of physical models is difficult to beat with a flat computer screen.

Furthermore, in the second year it is not possible to teach sophisticated energy modeling tools because students lack sufficient knowledge in environmental control systems. One strategy that I have been implementing to introduce more complex

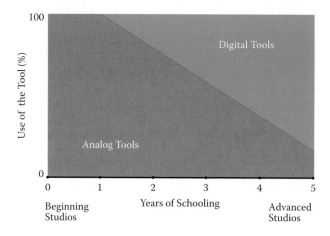

FIGURE 2.7 Use of analog and digital tools in studios.

Carbon-Neutral Architectural Design

concepts is to teach a specific portion of a program in an area of knowledge. An example is teaching them simple daylighting principles and then using the daylight module in Ecotect to calculate daylight factors inside their projects. Even though this method is not very precise because it is based on a manual method that uses a geometric version of the split flux method as outlined by the U.K. Building Research Establishment (BRE) and ignores sun position, orientation, and seasonal variations, it can provide quick and easy-to-understand results that have sufficient precision for second-year students. For more detailed analysis, performed by students in more advanced topics studios, Ecotect models are exported to Radiance, which provides a more physically accurate daylight simulation.

Another diagram illustrating the carbon-neutral design process in second-year architectural design studios was developed (Figure 2.8). The diagram follows the same structure as the general carbon-neutral design process in Figure 2.4, but in this case some of the general knowledge area rectangles now have additional rectangles inside them that represent specific strategies implemented in the studio in these areas of knowledge.

The diagram also indicates with a thicker outline, the areas in which digital tools are used. Four digital tools were used in second year: (1) Climate Consultant for climate analysis; (2) home energy efficient design (HEED) for energy modeling; (3) Ecotect for daylight analysis; and (4) PVWatts for renewable energy. Multiple sustainable concepts were taught and implemented in the projects, but only a few of them were evaluated with the digital tools indicated in the rectangles. As students

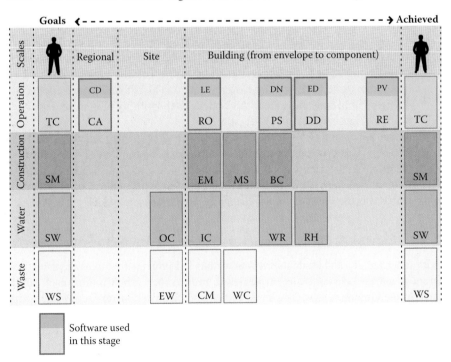

FIGURE 2.8 CNDP in second-year studio with areas of knowledge and strategies.

or architects progress through their project, they implement several of the strategies indicated in the diagram. The next paragraphs briefly discuss these strategies, which are also explained on the Carbon-Neutral Design Project site (aia.org/carbonneutraldesignproject) in the teaching section.

> **Climate design (CO).** Climate analysis consists of the survey of given climatic conditions to compare them with the requirements for human thermal comfort and to detect design opportunities that can be implemented to achieve comfort without the use of mechanical systems. Using climate analysis tools such as Climate Consultant (Milne, 2007) or Ecotect's Weather Tool, it is possible to propose building design strategies that can be implemented to achieve thermal comfort with reduced energy use (Figure 2.9). In topics studio traditional climate analysis was taken one step further to establish relationships between GHG emissions and climate zones.
>
> **Low-energy envelope (LE).** Design of the building envelope is critical because of its roles as a regulator of external conditions. Some projects were designed to operate only with passive systems, while in other projects mechanical heating was combined with passive heating and cooling strategies. Students used HEED (Milne, Gomez, LaRoche, and Morton, 2006) to determine energy use when mechanical systems were implemented and thermal comfort when only passive systems were used.
>
> **Efficient daylight (ED).** It is important to achieve a controlled distribution of light inside a space. Adequate illuminance levels should be provided while eliminating glare. Students had to design the fenestrations to achieve the necessary illuminance levels with daylight while minimizing glare. To achieve this they performed a simple daylighting analysis with Ecotect to determine if the shading system was also providing enough daylight inside the space (CIE standard overcast sky). This should ensure enough daylight but in constantly clear skies opens the possibility of glare and overheating (Figures 2.10, 2.11, 2.12, and 2.13).
>
> **Photovoltaic design (PV).** If the building was designed with good energy efficiency, it was possible to generate much of the required electrical energy on site. After students obtained the total building load with HEED, they used tools such as PVWatts to calculate and design the photovoltaic system, which was the type of renewable energy used in these projects.

2.3.2 Carbon-Neutral Architectural Design in Advanced Studios

With other faculty and students, the author has led projects that explore, using different scenarios, methods and strategies to reduce the carbon footprint of buildings. In the GreenKit project several studios and elective courses were taught with Michael Fox and Phyllis Nelson and funded by a grant from the U.S. Environmental Protection Agency (EPA) P3 program, as part of a student-driven research project titled "The Green Kit: A Modular, Variable Application System for Sustainable Cooling" (Yezell, Felton, La Roche, and Fox, 2007). This project was developed in

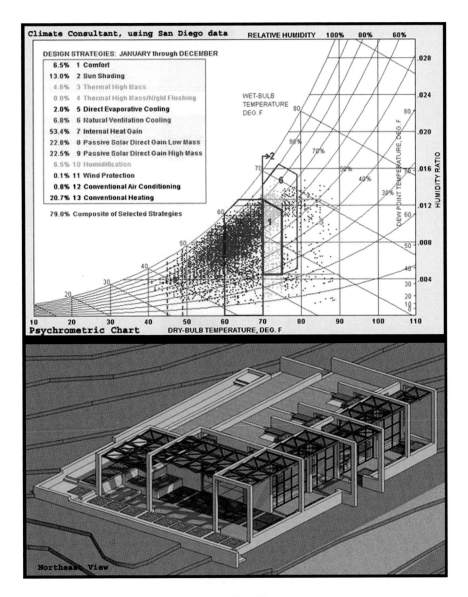

FIGURE 2.9 Climate design. (Student Erin Yazell.)

a joint effort of architecture and engineering students working with local organizations in several locations in the United States, Uganda, Mexico, Venezuela, and China (http://www.csupomona.edu/~p3team/). Students used computer tools to evaluate the performance of their ideas, even developing a rating system to evaluate different variables and select the best window system (Figures 2.14 and 2.15). Our rating system was based on the format of Malcolm Wells's original wilderness-based checklist for design and construction, which was revised by the Society of Building

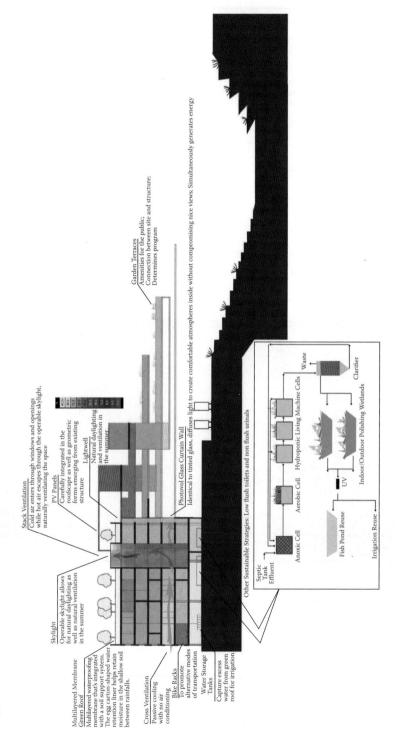

FIGURE 2.10 Daylight analysis in second-year student project. (Cal Poly Pomona students Kellene Kaas and Kenny Ngo.)

Carbon-Neutral Architectural Design

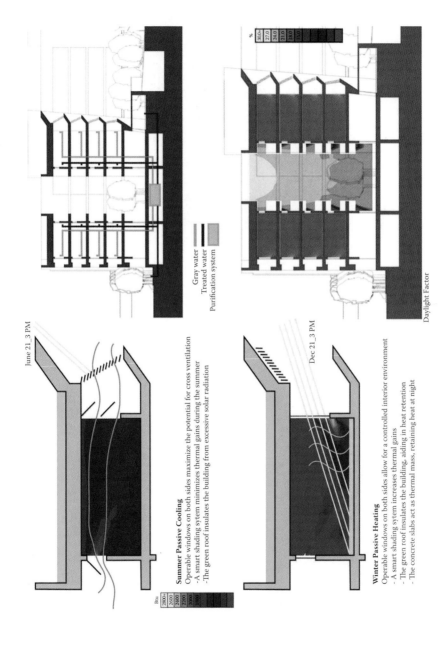

FIGURE 2.11 Daylight analysis in second-year student project. (Cal Poly Pomona students Jeremy Brunel and Ryan Dayag.)

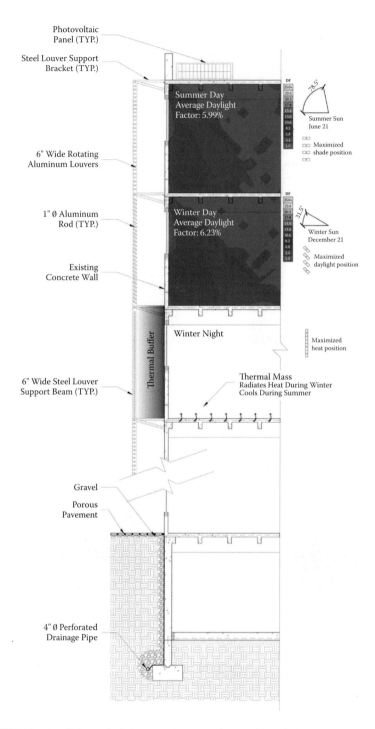

FIGURE 2.12 Daylight analysis in second-year student project. (Cal Poly Pomona students Sandeesh Sidhu and Greg Sagherian.)

Carbon-Neutral Architectural Design

FIGURE 2.13 Airflow analysis with wind tunnel. (Cal Poly-Pomona students Calvin Mensonides and Emily Tragish.)

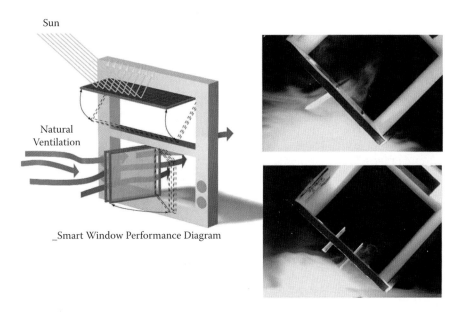

FIGURE 2.14 Green kit project window design.

Smart Window

Complete Smart Window

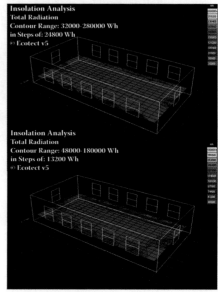
Ecotect Radiation Analysis

FIGURE 2.15 Green kit project window and daylight analysis.

Science Educators (SBSE) in 1999 and is available at http://www.sbse.org/resources/index.htm. A paper with our rating system was presented at the American Solar Energy National Conference in San Diego in 2008 (Epp, LaRoche, and Fox, 2008).

The Tijuana low-cost sustainable housing prototype is a continuing project under development with several Cal Poly Pomona faculty including Kyle Brown and Irma Ramirez, and many Cal Poly Pomona students. The goal was to design and build a low-cost, sustainable house, constructed from a variety of local waste materials for informal settlements in Tijuana. With this project we won the NCARB Grand Prize in 2008. The 35 m^2 prototype incorporates recycled, low-energy materials and systems in a very low-cost building. The walls are made out of papercrete, a mixture of recycled paper (newspaper in this case) and cement; the roof trusses are constructed from discarded wood pallets; all the windows were built by the students, some of them from 99-cent IKEA glass plates; and a very low-cost green roof tops the house in the living area. Students also developed several active systems for the house: a low-cost solar hot water system for domestic and radiant heating, a solar air heater for the living room, and even a low-cost PV system to power lights and a music system (Hansanuwat, Lyles, West, and La Roche, 2007; Jeerage, La Roche, and Spiegelhalter, 2008; West and La Roche, 2008). Students used computer modeling tools to predict the thermal performance, which was then compared with thermal data collected from the house.

In addition to these courses, the author has been teaching several carbon-neutral design studios. In 2007, architecture students proposed designs for carbon-neutral homes

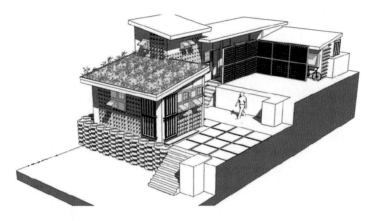

FIGURE 2.16 Tijuana prototype initial idea for layout and materials.

FIGURE 2.17 Tijuana prototype front exterior view in 2008.

in two Southern California climate zones: the hot and arid desert and the temperate coast. They developed low-energy buildings, integrating passive solar heating and cooling and renewable energy sources. In 2008 the second studio focused on developing two low-cost carbon-neutral housing prototypes, the first for Pomona with Habitat for Humanity and the second one for Tijuana, Mexico, with a nonprofit organization, CORAZON (Figures 2.16, 2.17, 2.18, and 2.19). The third studio in 2010 focused on affordable, sustainable, accessible housing for the U.S. Green Building Council (USGBC) competition in New Orleans. Some teaching strategies implemented in the studios were explained in the teaching section of the Carbon Neutral Design Project website (www.aia.org/carbonneutraldesignproject), and student projects are also posted on the website for ZeroCarbonDesign (http://www.zerocarbondesign.org). Recent studios in 2010 and 2011 focused on the design of carbon-neutral cities (Figure 2.20) or affordable housing for Pamo Valley in San Diego County.

FIGURE 2.18 Tijuana prototype side view in 2008.

FIGURE 2.19 Tijuana prototype interior view in 2008. (Photo by Tom Zasadzinsk.)

FIGURE 2.20 Oil platform city. (Cal Poly Pomona students Marcus Richeson, Alex Phung, Gabriela Barajas, and Ashi Martin.)

During the course of the studio, several exercises are usually developed, which involve the use of energy simulation tools and which will be discussed in the following sections. The exercises are organized in several steps that roughly correspond to design scales: regional, neighborhood, site, and building (including the skin and interior spaces; Figure 2.21). Students usually also participate in a 4-hour design charrette facilitated by Energy Design Resources and Southern California Edison that

Carbon-Neutral Architectural Design

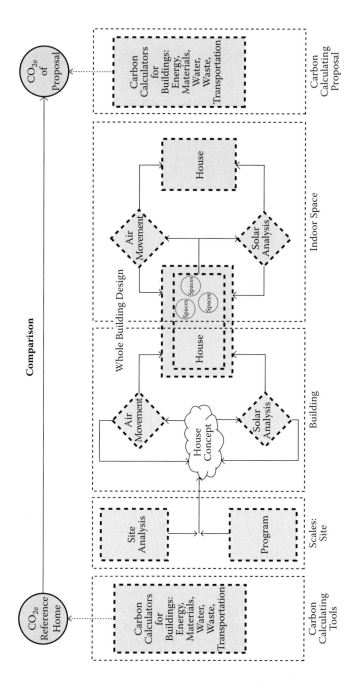

FIGURE 2.21 Topic studio design process.

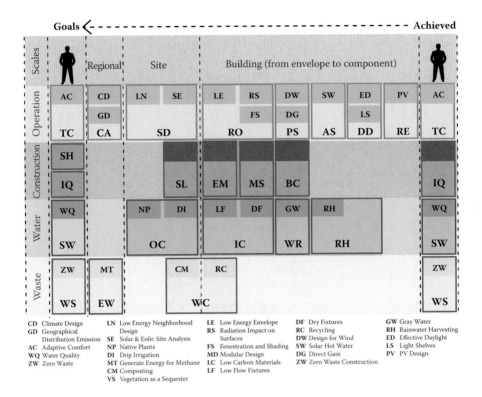

FIGURE 2.22 Carbon-neutral design process implemented in Advanced Studios.

simulates a real-life professional environment with role playing. During the charrette the students learn more about their roles as part of an interdisciplinary team in an integrated design process. This charrette has now been expanded to a 1 1/2 day event open to all Southern California schools of architecture.

Students are first introduced to the design process that they will follow during the quarter and the tools that they will use to achieve low-carbon buildings. A version adapted from Figure 2.4 of the general carbon-neutral design process was developed (Figure 2.22) in which the specific implemented strategies are explained. Many more strategies can be generated in each of the knowledge areas.

> **Climate design (CO).** Climate design is the most common strategy implemented during climate analysis. Weather files are analyzed and compared to thermal comfort requirements, and appropriate climate-responsive design strategies are proposed. Sometimes in these studios traditional climate analysis was taken one step further to try to understand relationships between GHG emissions and climate zones as expressed in the next section. This process is continued in the next strategy.
>
> **Geographical distribution of emissions (GD).** Students are not used to visualizing amounts of CO_2, and it is important that just as they have already learned how to visualize their projects in dimensional units of measurement,

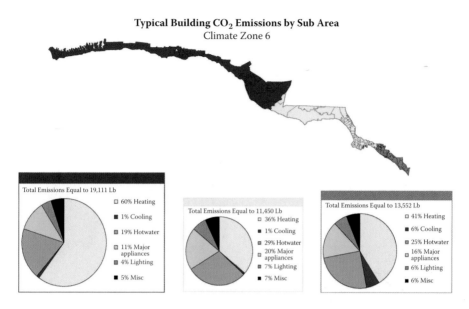

FIGURE 2.23 Geographical distribution of emissions.

they can also learn how to think in these units, that is, how much is a large quantity and how much is just a little bit. By quantifying the units in different climate zones, this exercise helps the students visualize units of CO_2 emitted per dwelling (Figure 2.23). The students determined annual residential CO_2 emissions for buildings according to their energy emissions and zip codes in different climate zones using a method developed in class that used Home Energy Saver (http://hes.lbl.gov) and census tract information (http://census.gov). This exercise helped them understand the relationship among carbon emissions, climate, and population density (Figure 2.24).

Solar and site analysis (SE). Building shading on outdoor spaces has an important effect on its perception and its use. If it is intended for winter use and it is shaded by a neighboring building, it will probably be dark, cold, and unused. Nor will it be used if it is intended for use during hot weather and it is not shaded. Ecotect was used to generate sun path diagrams in different outdoor locations to determine the effect of building shading on exterior spaces and their potential during the summer or winter. Because they indicate when radiation will be available, these can also be used to determine performance of PV systems or direct gain systems for passive and active heating. Adjustments in building massing can then be proposed to improve shading or solar access. Figure 2.25 shows how a group of students developed and tested a shading system for a studio project in Palm Springs, CA. The second sun path diagram shows how the shading system reduces the number of hours in which the shared outdoor space receives direct sun. Figure 2.26 shows how the sun path diagram can be used as a tool to determine solar exposure in different locations at the site, helping to determine if it will be uncomfortable during different seasons.

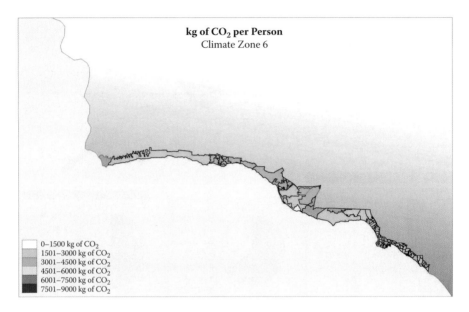

FIGURE 2.24 Geographical distribution of emissions.

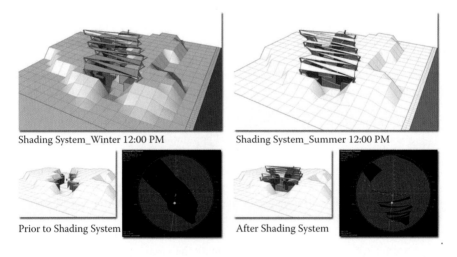

FIGURE 2.25 Solar and site analysis. (Cal Poly Pomona students Brandon Gulloti, Jon Gayomali, and Garret Van Leuween.)

Low-energy neighborhood design (LN). Students work in groups to propose a sustainable neighborhood plan in each climate zone, implementing concepts that would reduce the ecological footprint of the neighborhood and the city. These concepts include control of solar radiation and air movement, reduction of water use and resources, and community well-being. Figure 2.27 shows how the buildings are grouped to maximize southern exposure and how the sun path diagram is again used to analyze shading

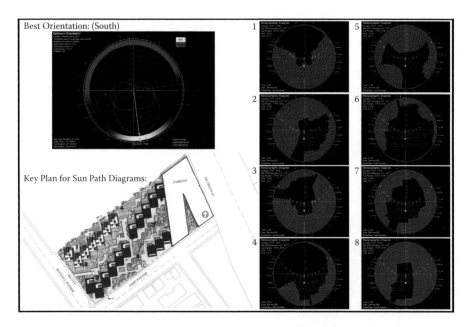

FIGURE 2.26 Solar and site analysis. (Cal Poly Pomona students Serge Mayer and Ryan Cook.)

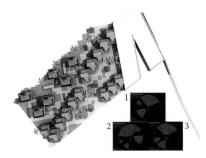

FIGURE 2.27 Low-energy neighborhood proposal. (Cal Poly Pomona students Ryan Hansanuwat, Marcos Garcia, and John Duong.)

in three locations at the site. Cross-ventilation is also promoted throughout the houses.

Low-energy envelope (LE). As in the beginning years, some projects were designed to operate only with passive systems, whereas in other projects, mechanical heating was combined with passive heating and cooling strategies. Students used HEED (Milne, 2001) to determine energy use when mechanical systems were implemented and thermal comfort when only passive systems were used. GHG emissions from energy use were determined by multiplying the energy use calculated with HEED by a conversion factor of 0.62 kg CO_2e/kWh for electricity, the average value for the United States. Students also had to develop concept diagrams to explain the energy efficient features of the building (Figures 2.28, 2.29, and 2.30).

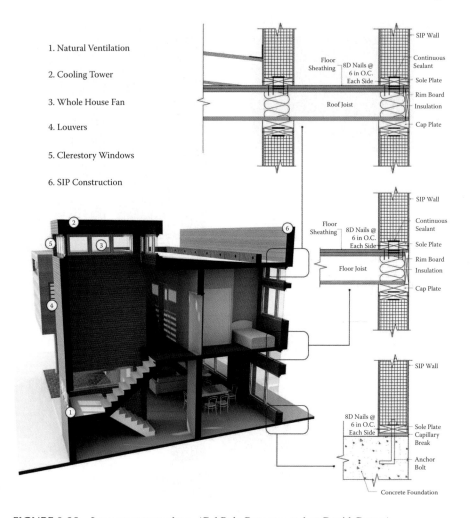

FIGURE 2.28 Low-energy envelope. (Cal Poly Pomona student David Castro.)

Radiation impact on surfaces (RS). Building surfaces receive varying amounts of solar radiation depending on the climate and latitude of the site and the orientation and tilt of the surface. Incident solar radiation can have a significant impact on buildings by affecting the surface temperature of opaque materials and thus increasing the rate of heat transfer by conduction or by radiation and conduction through windows. Surfaces that receive more solar radiation might require additional shading during the summer or would be suitable for solar hot water collectors or photovoltaic panels or to place a window to provide solar gains to the interior of the building. Ecotect's option for solar access is used to calculate and analyze irradiation on the exterior surfaces of the building (Figures 2.31, 2.32, and 2.33). The three-dimensional (3-D) models describe the incident radiation on external surfaces, and the colors describe different amounts of solar radiation.

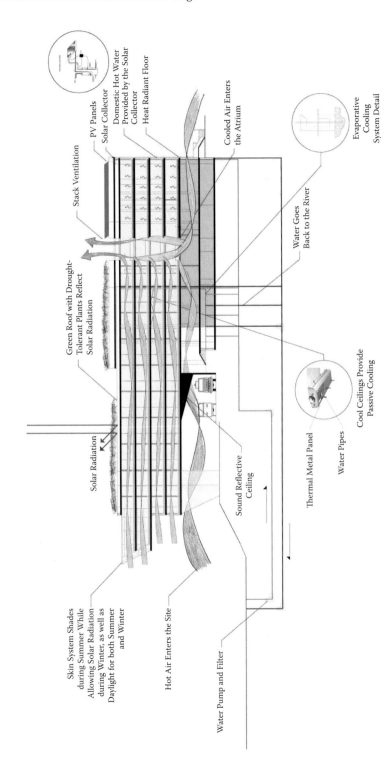

FIGURE 2.29 Low-energy envelope. (Cal Poly Pomona students Emmanuele Gonzalez and Tyler Tucker.)

Sustainable Strategies

1 Stereographic Diagram (showing the effects of the overhang)
2 Photovoltaic Panels - LEED EA 10 (to be installed in the future)
3 Wood Siding with Non-toxic Paint
4 Energy Star Appliances
5 Wood Louvers Made of Local Materials LEED MR 2.2
6 Concrete Pavers - LEED SS 4.1
7 Siphonic Roof Drains (for the water catchment system)
8 Solar Flat Plate Collectors (for the solar water heater)
9 Low-Flush Toilets
10 Native Plants - LEED SS 2.4
11 Double-Glazed Low-E Windows LEED EA 4.2
12 Debri from Local Demolitions (used to build up the ramp)
13 Porous Asphalt - LEED SS 4.1

FIGURE 2.30 Low-energy envelope. (Cal Poly Pomona student Leslie Cervantes.)

Figure 2.31 studies the interactions between solar radiation and building massing in a hot and dry climate, while Figures 2.32 and 2.33 study the effects of a milder climate in Los Angeles, CA.

Fenestration and shading (FS). Solar radiation entering through windows can contribute a significant amount of heat into the building, which can be beneficial when heating is required but can also be a liability when thermal gains must be blocked and cooling is needed. Radiation can be regulated with properly designed shading systems. Daylighting, however, should be provided during the whole year. The objective of this exercise is to design and optimize an integrated window/shading system to provide the necessary solar protection in the summer and solar radiation in the winter (if required) while providing daylight. The first step in this process is to analyze weather files using the psychrometric chart from Climate Consultant to determine when solar radiation is needed inside the space. The sun-shading chart from Climate Consultant plots solar position (Szokolay, 1996) with outdoor temperature and dates and times in one diagram and indicates when shade is needed. Inside, Ecotect students used the option for solar access analysis and selected the option to determine the incident solar radiation, while a grid inside the space shows irradiation values (Figures 2.34, 2.35, 2.36, 2.37, and 2.38). The sun path diagram can also be used to determine shading provided by exterior elements on windows or when radiation will penetrate into a space through the different openings (Figure 2.39).

Carbon-Neutral Architectural Design

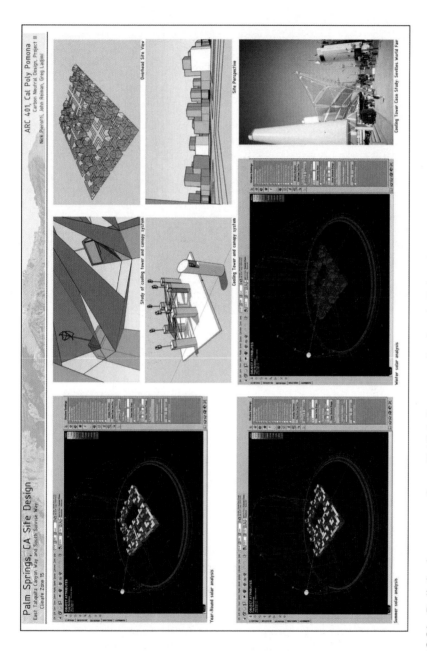

FIGURE 2.31 Radiation impact on surfaces. (Cal Poly Pomona student Jonatha Reiman.)

FIGURE 2.32 Radiation impact on surfaces. (Cal Poly Pomona student Serge Mayer and Ryan Cook.)

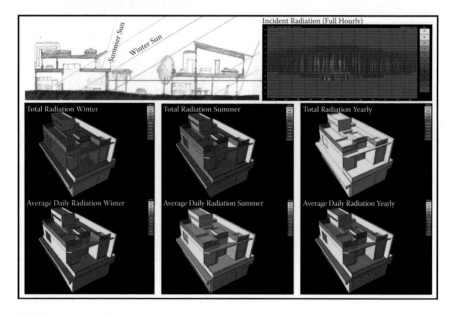

FIGURE 2.33 Radiation impact on surfaces. (Cal Poly Pomona student Ryan Cook.)

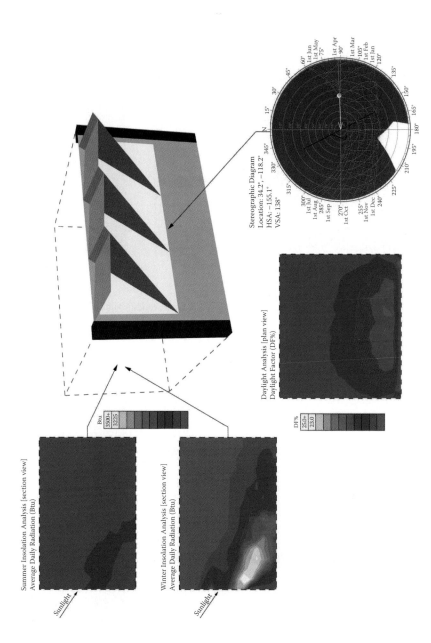

FIGURE 2.34 Fenestration and shading. (Cal Poly Pomona students Emmanuele Gonzalez, Tyler Tucker, and Rogelio Diaz.)

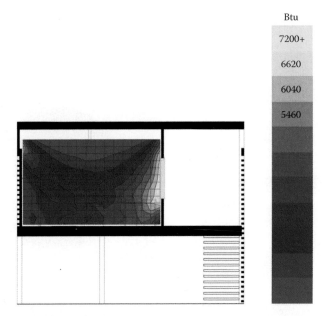

FIGURE 2.35 Fenestration and shading. (Cal Poly Pomona student Nancy Park.)

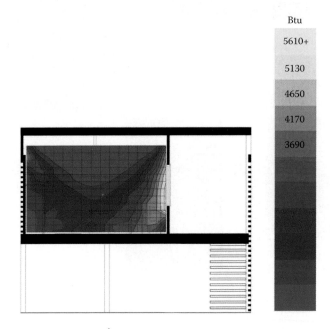

FIGURE 2.36 Fenestration and shading. (Cal Poly Pomona student Nancy Park.)

Carbon-Neutral Architectural Design

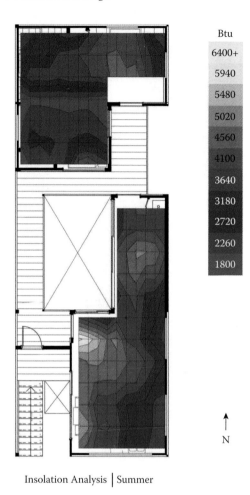

Insolation Analysis | Summer

FIGURE 2.37 Fenestration and shading. (Cal Poly Pomona student Nancy Park.)

Efficient design with daylight (ED). It is important to achieve a controlled distribution of light inside a space. Adequate illuminance levels should be provided while eliminating glare. Daylighting can be provided by sidelighting and toplighting. Sidelighting admits light from the perimeter walls of a building and can provide daylight in up to 70% of the footprint. Sidelighting is more effective when the ceiling is above 8 feet high and closer to 10 feet high. Toplighting admits light through the top of a building through vertical or horizontal glazing such as roof monitors, clerestories, sawtooth elements, skylights, and atria. Students had to design fenestrations to achieve the required illuminance levels with daylight while minimizing glare. They used Ecotect and Radiance to determine illuminance and luminance levels inside the house. Students first did a simple daylighting analysis using Ecotect to indicate if the shading system was also providing enough daylight inside the space (CIE standard overcast sky). This analysis ensures that

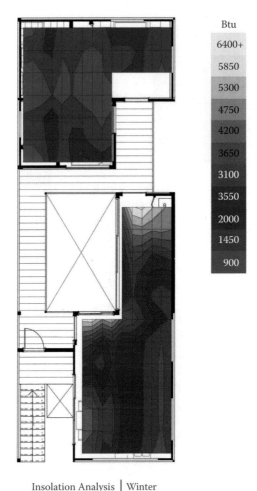

Insolation Analysis | Winter

FIGURE 2.38 Fenestration and shading. (Cal Poly Pomona student Nancy Park.)

there is enough daylight but in constantly clear skies opens up the possibility of glare and overheating. In addition to these analyses, the students in these studios generated additional, more precise calculations with Ecotect and Radiance. The model data are exported to Radiance, and students used contour lines to show illuminance and luminance values. Radiance provides the option to use weather files and generates results for specific days and hours (Figures 2.40, 2.41, 2.42, 2.43, 2.44). Figures 2.40, 2.41, and 2.42 show luminance analysis inside spaces during different hours and seasons, while Figures 2.43 and 2.44 show illuminance values inside the spaces.

Light shelves (LS). Light shelves bounce visible light up toward the ceiling, which then reflects it down deeper into the interior of a room. A horizontal surface will reflect the light with an angle equal to the angle of incidence. A highly reflective surface will reflect the light with an angle equal to the

Carbon-Neutral Architectural Design

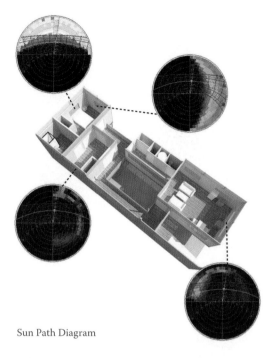

Sun Path Diagram

FIGURE 2.39 Fenestration and shading. (Cal Poly Pomona student Nancy Park.)

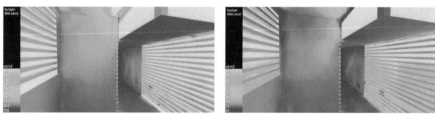

Luminance: Summer   Luminance: Winter

FIGURE 2.40 Design with daylight. (Cal Poly Pomona student Jillian Schroetinger.)

angle of incidence of the incoming daylight. If the light shelf is painted with a rougher material, then it will provide a more diffuse spread of light, which will scatter the light with varied angles and will provide a more even coverage of the ceiling. If it is finished with a smooth surface then it bounces deeper inside the space. The ceiling should be painted with a light color to reflect more the light down into the area underneath.

Design with wind (DW). Natural ventilation can be a very effective cooling strategy. In hot and humid climates it can provide cooling to our bodies, while in hot and dry climates it can cool the interior structure of the building during the night, which then acts as a heat sink during the day. To adequately implement natural ventilation it is necessary to determine how the air flows through the openings and spaces of the building. Using

FIGURE 2.41 Design with daylight. (Cal Poly Pomona student Charles Campanella.)

FIGURE 2.42 Design with daylight. (Cal Poly Pomona students Aireen Batungbatal and Alexandra Hernandez.)

FIGURE 2.43 Design with daylight. (Cal Poly Pomona student Charles Campanella.)

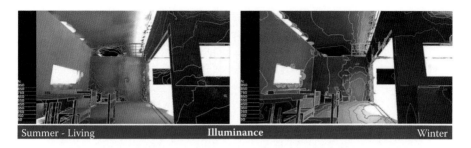

FIGURE 2.44 Design with daylight. (Cal Poly Pomona students Aireen Batungbatal and Alexandra Hernandez.)

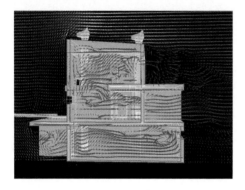

FIGURE 2.45 Design with wind. (Cal Poly Pomona student Jonathan Reiman.)

Ecotect and Win Air, students determined the pressures and direction of air movement through different areas of the building. Layers had to be cut horizontally and vertically to understand airflow patterns. Students also used "class-made" wind tunnels in second-year studio and in the GreenKit project. Even though these are very simple computational fluid dynamics (CFD), analysis that does not consider thermal buoyancy effects permits students to go beyond the "magic arrow" type analysis, and they serve to validate the airflow concepts sketched out by the students (Figures 2.45, 2.46, and 2.47).

Direct gain (DG). The simplest passive solar heating system is a direct gain system. It consists of an insulated space with equatorial-facing glazing for solar gains that are absorbed by thermal storage mass and reemitted later. Thus, the living space where the mass has been stored is a "live-in" collector. These systems require a large equatorial-facing glazing with the living space directly behind and an exposed thermal mass in the floor, ceiling, or walls, sized with enough capacity for thermal storage and positioned for solar exposure. A direct gain system will use 60–75% of the sun's energy striking the windows.

Solar hot water (SW). Active solar is used to designate mechanical devices whose sole purpose is to collect solar energy in the form of heat and then to

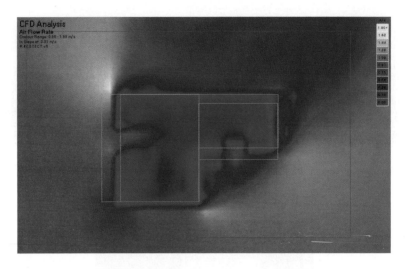

FIGURE 2.46 Design with wind, analysis of the Tijuana House.

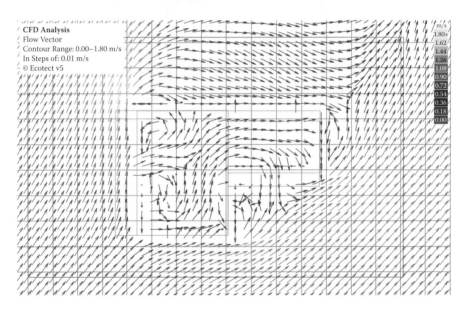

FIGURE 2.47 Design with wind, analysis of the Tijuana House.

store it for later use. The working fluid is usually water pumped from the collector to the storage tank. Solar heating collectors can convert over 80% of the sunlight that hits them into useable heat. The main applications of solar thermal heating are domestic water heating (DWH), space heating, and direct solar pool heating. Domestic water heating is an area of high potential for the application of solar energy in homes because it is useful year-round. There are many types of systems, and they differ in how the transfer fluid is circulated between collector and storage tank and the freeze protection provided.

Carbon-Neutral Architectural Design

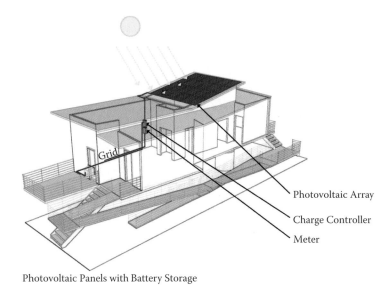

Photovoltaic Panels with Battery Storage

FIGURE 2.48 Renewable energy. (Cal Poly Pomona student Jorge Medrano.)

Photovoltaic systems (PV). Students calculated the total building load with HEED and then used tools such as PVWatts to calculate and design the photovoltaic system, which was the type of renewable energy used in these projects (Figure 2.48). A PV array produces power when exposed to sunlight and has a number of other components to properly conduct, control, convert, distribute, and store the energy produced by the array.

Modular design (MD). Modulation in architecture is important to maximize structural and functional efficiency with minimum use of materials. Modular design is also important when designing a building for deconstruction or adaptation to another function and provides a building with more flexibility.

Low-carbon materials (LC). GHG emissions from construction occur during the fabrication and transportation of materials to the building and the construction of the building. They are more difficult to calculate because it is not easy to determine the amount of the different materials in the building, their origin and manufacturing process, and the transportation method and distance from origin to the site. Materials with little embodied energy (and thus little embodied GHG) should be used. More nonrenewable energy to produce the material means that more carbon will be embodied in it. The transportation of the material to the construction site will also affect emissions, and materials that must be transported a long distance will usually have higher values of embodied emissions. Students used buildcarbonneutral (http://buildcarbonneutral.org/) or Athena Institute's Eco Calculator (http://www.athenasmi.org/tools/ecoCalculator/index.html) to estimate the total emissions for the building. Emissions were later divided by the estimated life of the building (50 years was used) to obtain the emissions per

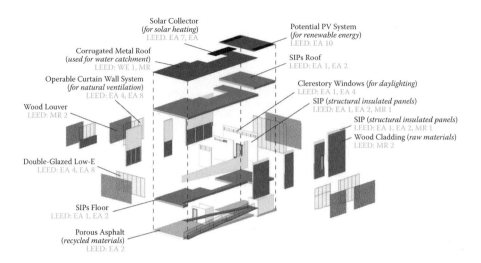

FIGURE 2.49 Material selection and building integrated sustainable strategies. (Cal Poly Pomona student Leslie Cervantes.)

year in kg CO_2e/year. Some of these materials can also be used in different sustainable strategies (Figure 2.49).

Zero-waste construction (ZW). This refers to optimization of the construction process to reduce construction waste and solid waste. Greenhouse gases are also produced by the construction machinery and the transportation of supplies, materials, and people to the construction site. This process is usually out of the scope of the architectural design studios.

Native plants (NP). Native plants save water because they are accustomed to local conditions including temperature and rainfall. They also need little long-term maintenance after plantings are established.

Drip irrigation (DI). Drip irrigation is the gradual application of water to the plants' root zone. Drip irrigation is more efficient than watering the surface because water is not wasted on nongrowth areas, and the plant root zone is maintained at its ideal moisture level. Drip irrigation maintains an optimum moisture level in the soil while less water is lost to the sun and the wind compared to spray-based irrigation.

Low-flow fixtures (LF). Potable water is a precious resource that is available for human consumption in very limited quantities. Furthermore, water consumed in buildings also generates CO_2 emissions because it usually takes energy to move and treat. The farther this water must be transported and treated for human consumption, the more energy is used. If this energy is from nonrenewable sources, then it will generate emissions. Students estimated water consumption in the building and then implemented measures to reduce its consumption inside the building through various conservation measures that would prevent waste of this resource. Methods to reduce potable water use include installing sensors and flow restrictors on water fixtures.

Dry fixtures (DF). A more drastic method to reduce the use of water in buildings is installing dry fixtures such as composting toilets and waterless urinals; however, these do not usually affect the design of the building.

Gray water reuse (GW). Gray water has already been used in the building but does not contain harmful pathogens. It is typically defined as the wastewater produced from clothes washers, baths and showers, kitchen sinks, dishwashers, and lavatories. In some municipalities, only wastewater from clothes washers is permitted to be used as gray water. If properly collected and stored, gray water can be safely reused, reducing consumption of potable water and the load on septic tanks and leach fields. The wastewater generated by toilets is called black water.

Rainwater harvesting (RH). Harvested rainwater is captured from the roofs of buildings and can be used for indoor needs at a residence, irrigation, or both. Harvesting rainwater reduces the consumption of treated potable water, thus reducing a building's effect on the environment and its carbon footprint. Figures 2.50, 2.51, and 2.52 show some student proposals for New Orleans with differet types of rainwater harvesting systems.

Generate energy from methane (MT). Waste in a landfill generates methane, which can be used to generate energy. The EPA requires all large landfills to install collection systems to minimize the release of methane. These gases are collected by drilling "wells" into the landfills and collecting the gases through pipes and combining them with natural gas to fuel conventional combustion turbines or used to fuel small combustion or combined cycle turbines. Landfill gas can also be used in fuel cell technologies, which use chemical reactions to create electricity and are much more efficient than combustion turbines.

Composting (CM). Composting is the aerobic, biological decomposition of the organic constituents of wastes under controlled conditions. Some heat is liberated during the process, causing the compost to warm. During aerobic composting, microorganisms consume organic material; nitrogen is used by these microorganisms for metabolism and cell growth, while carbon supplies an energy source. Some materials such as kitchen scraps provide nitrogen, while other materials such as leaves and dried straw provide carbon. The rate of composting depends on a number of factors, but those of major importance are moisture content, carbon-to-nitrogen ratio, aeration, temperature, and microbial population. The end product of composting is a dark, crumbly substance. Because the volume of the material is reduced to a quarter of its original size, composting is an excellent way to recycle yard and kitchen wastes while reducing the volume of waste sent to landfills. In addition to these benefits, the energy that is produced as a by-product of the composting process can be used to heat a space. Student Perry's tests at the Lyle Center for Regenerative Studies at Cal Poly demonstrated that it is possible to heat a space with the waste heat generated during the composting process. During her tests in which she monitored for a period of 30 days, a 3-gallon container reached a temperature of 160°F and an average of 118°F. Perry estimated her small composting pile to generate 2,500 Btu/hr of heat at peak capacity.

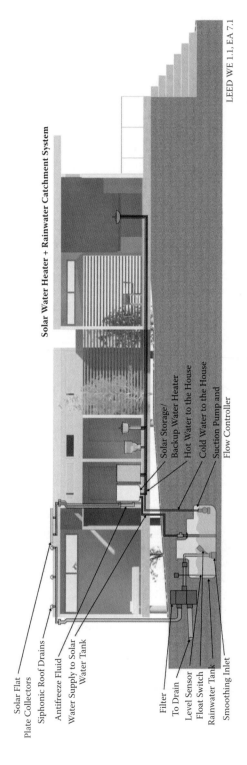

FIGURE 2.50 Rainwater harvesting. (Cal Poly Pomona student Jillian Schrotinger.)

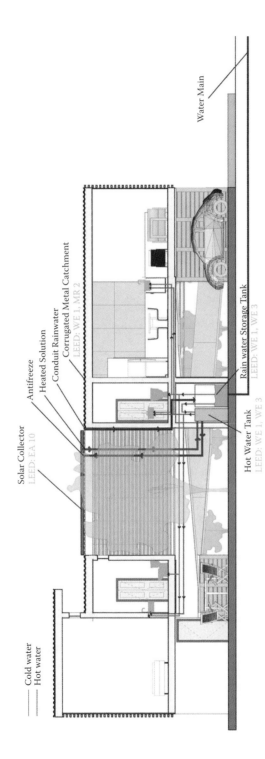

FIGURE 2.51 Rainwater harvesting. (Cal Poly Pomona student Leslie Cervantes.)

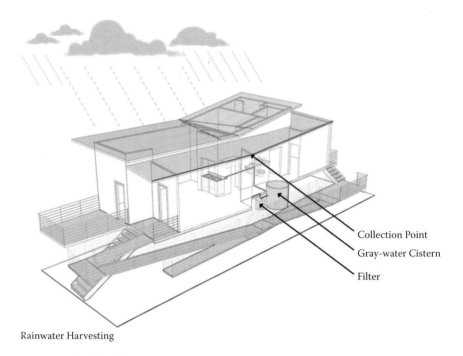

Rainwater Harvesting

FIGURE 2.52 Rainwater harvesting system. (Cal Poly Pomona student Jorge Medrano.)

Recycling (RC). Recycling reduces the amount of solid waste that we generate and the amount of greenhouse gases. According to the EPA (2011) recycling reduces emissions from energy consumption, reduces emissions from incinerators, reduces methane emissions from landfills, and increases storage of carbon in trees. The EPA estimates that increasing the U.S. recycling rate from its current level of 27% to 35% would reduce greenhouse gas emissions by 11.4 million metric tons of carbon equivalent over landfilling the same material.

2.4 INTEGRATION

All of these strategies are part of an integrated design process in which the designer implements a set of strategies to reduce the carbon footprint of a building in the areas of energy, waste, water, and construction. An example is the Pamo Valley project in which students from the department of architecture at Cal Poly Pomona are working with HMC Architects to design and build two city-owned homes in Pamo Valley that were destroyed in the 2007 California wildfires. These homes will serve as sustainable low-cost alternatives to Federal Emergency Management Agency (FEMA) prototypes, adapted to local conditions, and with the potential to become replicable alternatives for fire-prone areas in Southern California.

Students in this studio have developed the initial design of these two houses. In addition to replicability, fire resistance, adaptation to the environment, and cost, the

Carbon-Neutral Architectural Design

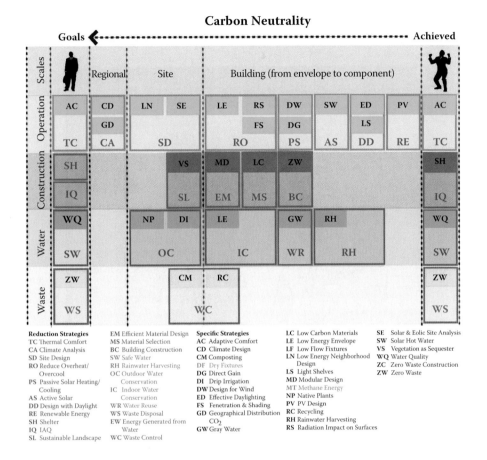

FIGURE 2.53 Carbon-neutral design diagram. (Team Brandon Ro and Aaron Locke.)

students had to reduce the project's carbon footprint and energy use intensity. To achieve this, they implemented the carbon-neutral design process described in this chapter in the areas of energy, waste, water, and construction. Figure 2.53 shows the strategies implemented by Team Ro–Locke, while Figure 2.54 shows where they are implemented in the project. Figure 2.55 shows the strategies by Team Tolios–Behbahani and Figure 2.56 shows where they are implemented in the house.

2.5 CARBON-NEUTRAL ARCHITECTURAL DESIGN PROCESS IN PRACTICE

Implementing this method in architectural education prepares future architects but also means that it will take several years to impact the built environment after they graduate and begin to practice architecture. If this CNDP is adopted by current practitioners, the process will have a more immediate impact in the built environment. This CNDP has begun to be implemented in HMC Architects in California as part of

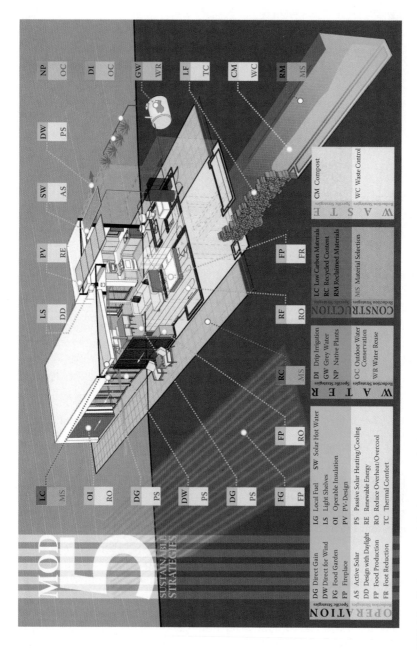

FIGURE 2.54 Carbon-neutral design strategies in project. (Brandon Ro and Aaron Locke.)

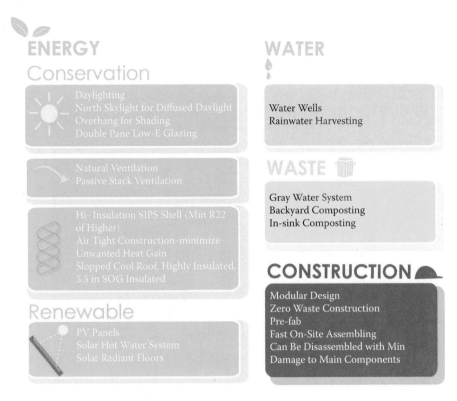

FIGURE 2.55 Carbon-neutral design diagram. Team Dimitrios Tolios and Parinaz Behbahani.

its high-performance architecture initiative. Strategies to achieve carbon neutrality are being implemented in several projects depending on programmatic and site conditions. Three projects in which this method, still under development, is being implemented in the initial design phases of an architectural design process are presented.

The first project is a school building ideas competition for the Los Angeles Unified School District. Our goal was to design a building with a very small carbon footprint, as close as possible to zero. To achieve this goal, different strategies were implemented and are presented in Figure 2.57. Figures 2.58, 2.59, 2.60, 2.61, 2.62, 2.63, and 2.64, produced by E. Carbonnier, indicate where these strategies can be implemented in the building, while Figure 2.64 compares the emissions in each of the four areas with those of an equivalent size school building. Even though this project was not selected by the Los Angeles Unified School District (LAUSD), it will probably be developed further and adapted for another location in Southern California.

Our proposal for a new classroom building for California State University at Monterey Bay includes the implementation of this method to reduce the building's carbon footprint. This building will house the School of Business and the School of Information Technology and Communication Design (ITCD) and includes 5600 m^2 of classroom, laboratory space, and offices. Among other things, the new classroom buildings' shape is a response to the sun and wind to form sheltered entries,

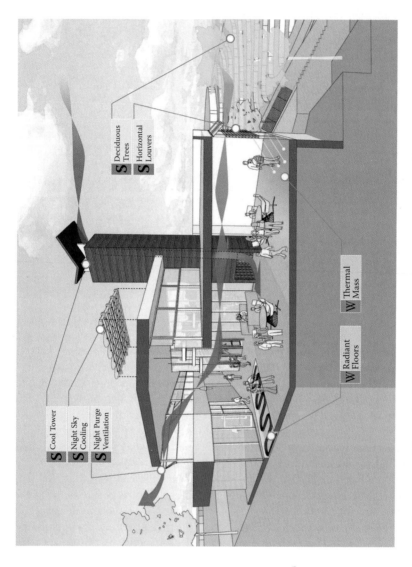

FIGURE 2.56 Preliminary proposal for academic building for Cal Poly Pomona Collins School of Management. (Courtesy of HMC Archlab.)

Carbon-Neutral Architectural Design

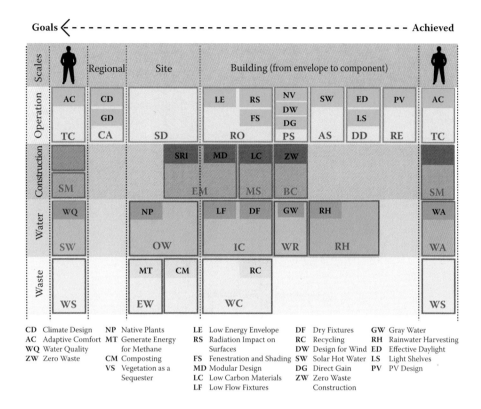

FIGURE 2.57 Strategies in the carbon-neutral design process for flex school building.

courtyards, and sunny urban spaces. Some of the strategies that are being implemented include appropriate fenestration and shading, natural and displacement ventilation (depending on the space), effective daylight design, solar hot water, etc. (Figure 2.65).

We are also beginning work on a new building for the Collins College of Hospitality Management at Cal Poly Pomona University. The 1500 m² facility for student services, classrooms, and administrative offices is designed to augment those currently located in the College and will house administrative offices for instructional staff and student services, such as lecture rooms and classrooms, in a low carbon building. Figure 2.66 illustrates an initial concept diagram with some design strategies that will help to achieve this, such as night ventilation, a cool tower, radiant cooling and shade.

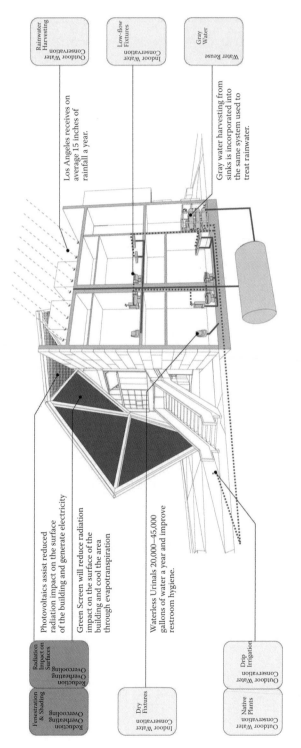

FIGURE 2.58 Water and energy strategies for flex school building. (Courtesy of HMC Archlab.)

Carbon-Neutral Architectural Design

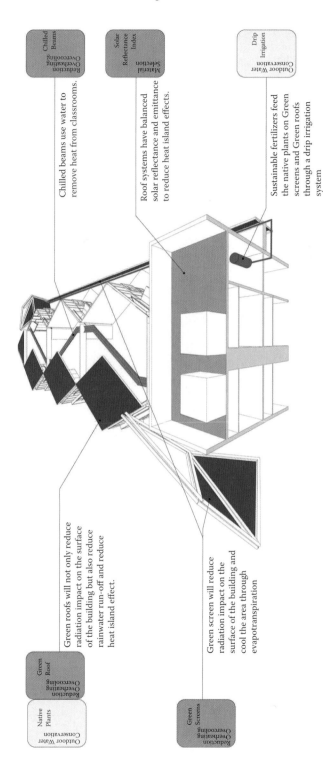

FIGURE 2.59 Water and energy strategies for flex school building. (Courtesy of HMC Archlab.)

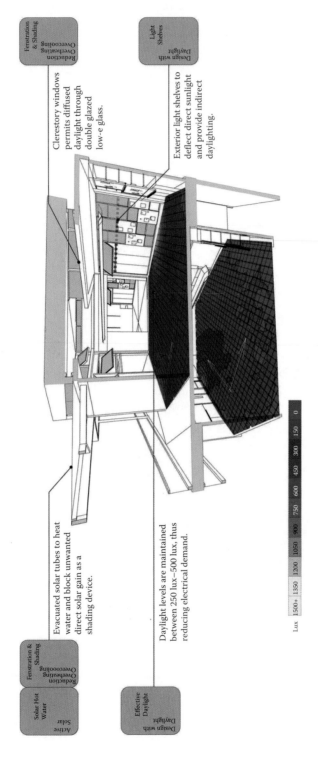

FIGURE 2.60 Energy strategies for flex school building. (Courtesy of HMC Archlab.)

Carbon-Neutral Architectural Design

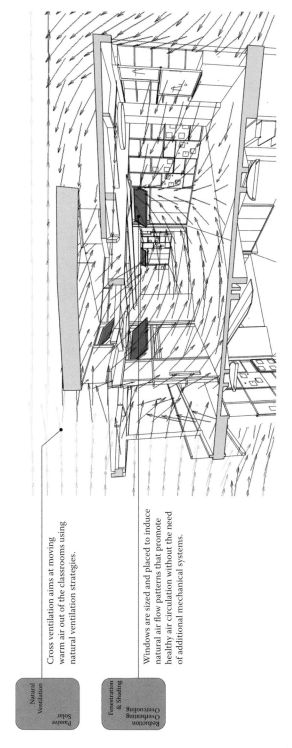

Natural Ventilation — Passive Solar
Cross ventilation aims at moving warm air out of the classrooms using natural ventilation strategies.

Fenestration & Shading — Reduction Overheating Overcooling
Windows are sized and placed to induce natural air flow patterns that promote healthy air circulation without the need of additional mechanical systems.

FIGURE 2.61 Energy strategies for flex school building. (Courtesy of HMC Archlab.)

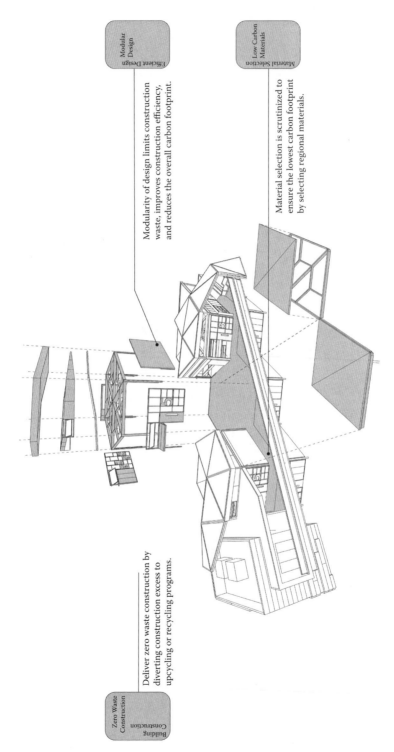

FIGURE 2.62 Construction strategies for flex school building. (Courtesy of HMC Archlab.)

Carbon-Neutral Architectural Design

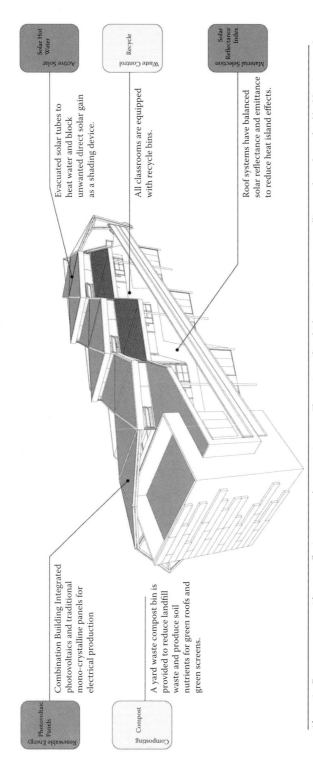

FIGURE 2.63 Energy and waste strategies for flex school building. (Courtesy of HMC Archlab.)

Alternative Energy systems provide a cost effective way to reduce operation costs. The project utilizes high generating mono-crystalline panels on the roof and building integrated photovoltaics on the triangulated shading system as it wraps around the roof perimeter. Hot water is generated by using evacuated solar tubes and a drainback system. The solar hot water system also serves as a horizontal overhang block unwanted solar direct gain.

Carbon Footprint

A carbon footprint is a commonly used unit of measurement that measures human impact on the environment as it relates to climate change. The unit of measurement is carbon dioxide equivalent (CO_2e) and represents all greenhouse gases that are emitted in day-to-day operations. The lower the unit, the less environmental degradation is committed.

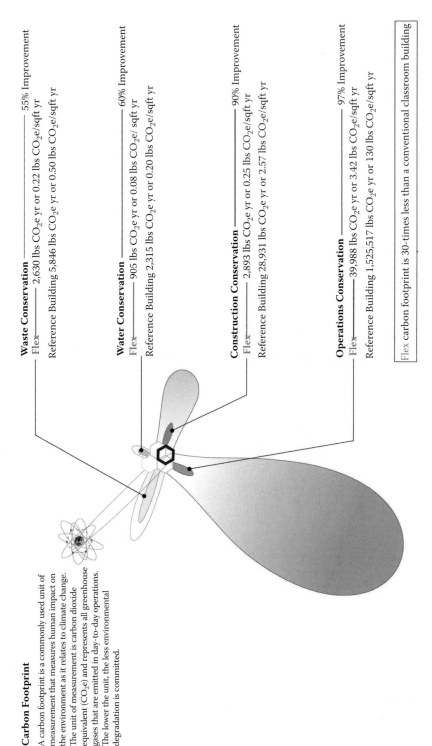

Waste Conservation ———————— 55% Improvement
Flex ——— 2,630 lbs CO_2e yr or 0.22 lbs CO_2e/sqft yr
Reference Building 5,846 lbs CO_2e yr or 0.50 lbs CO_2e/sqft yr

Water Conservation ———————— 60% Improvement
Flex ——— 905 lbs CO_2e yr or 0.08 lbs CO_2e/sqft yr
Reference Building 2,315 lbs CO_2e yr or 0.20 lbs CO_2e/sqft yr

Construction Conservation ———————— 90% Improvement
Flex ——— 2,893 lbs CO_2e yr or 0.25 lbs CO_2e/sqft yr
Reference Building 28,931 lbs CO_2e yr or 2.57 lbs CO_2e/sqft yr

Operations Conservation ———————— 97% Improvement
Flex ——— 39,988 lbs CO_2e yr or 3.42 lbs CO_2e/sqft yr
Reference Building 1,525,517 lbs CO_2e yr or 130 lbs CO_2e/sqft yr

Flex carbon footprint is 30-times less than a conventional classroom building

FIGURE 2.64 Comparison of emissions for flex school building. (Courtesy of HMC Archlab.)

Carbon-Neutral Architectural Design

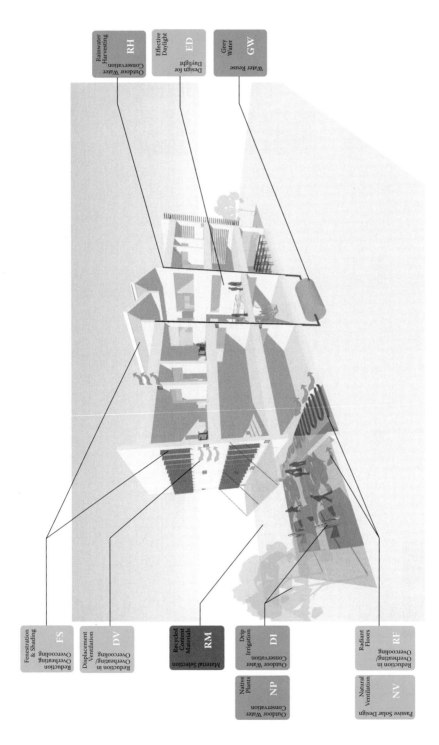

FIGURE 2.65 Some initial water, construction, and energy strategies in the new CSUMB classroom building. (Courtesy of HMC Archlab.)

FIGURE 2.66 Strategies in the carbon-neutral design process for CSUMB classroom building.

3 Thermal Comfort

Carbon-neutral buildings should provide thermal comfort with minimum energy use and minimum greenhouse gas (GHG) emissions. In the same way that a building that provides an adequate thermal environment with excessive emissions is unethical, a building that reduces emissions at the expense of an uncomfortable and unhealthy environment is unsuccessful. Thus, it is necessary to have a basic understanding of human thermal comfort to achieve a low-carbon, healthy, and thermally comfortable building that is environmentally friendly.

3.1 PSYCHROMETRICS

Clean air is a mixture of 10 gases: nitrogen, oxygen, argon, carbon dioxide, neon, helium, methane (CH_4), krypton, hydrogen, and xenon. Five of these travel alone as atoms, while the rest have atoms bonded together forming molecules. About 78.09% of the air is made out of nitrogen atoms, 20.95% is oxygen, and argon is about 0.53%. Carbon dioxide is about 0.04% (392.39 ppm in 2010; NOAA, 2010) and rising continually. Psychrometry is the science that deals with the study of this mixture, essentially the relationship of these gases with water vapor. This is important because the amount of water vapor in the air affects thermal comfort. The graphic representation of these relationships is the psychrometric chart. Figure 3.1 shows the relationship of these variables with each other in the American Society of Heating, Refrigerating, and Air-Conditioning Engineers' (ASHRAE's) psychrometric chart at sea level.

The psychrometric chart allows us to understand several relationships between concepts. For architectural design, the most important of these concepts to understand are the dry-bulb temperature, the humidity ratio, the relative humidity, and enthalpy. These concepts are shown in a simplified version of the psychrometric chart (Figure 3.2).

The dry-bulb temperature indicates the amount of heat in the air and is indicated on the horizontal axis (Figure 3.3). For molecules of air, the temperature is directly proportional to the mean kinetic energy of the air molecules. It is usually measured in degrees Celsius, Kelvin, or Farenheit (°C, K, or °F).

The moisture content or absolute humidity indicates the amount of moisture by weight within a given weight of air and is indicated on the vertical axis (Figure 3.4). It can be measured in lbs/lb, kg/kg, or g/kg (units of moisture per unit of dry air). Vapor pressure is linearly related to absolute humidity, so it can also be shown on this scale.

The relative humidity expresses the relationship between the amount of water vapor in the air and the maximum it can hold at that specific temperature and is expressed as a percentage (%). The relative humidity is indicated by curves that converge toward the left side of the psychrometric chart (Figure 3.5). The curves

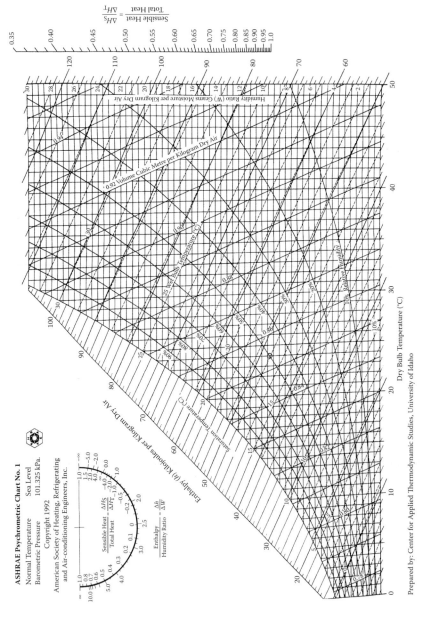

FIGURE 3.1 Psychrometric chart. (From ASHRAE, *Handbook of Fundamentals*, 2005. Used with permission.)

Thermal Comfort

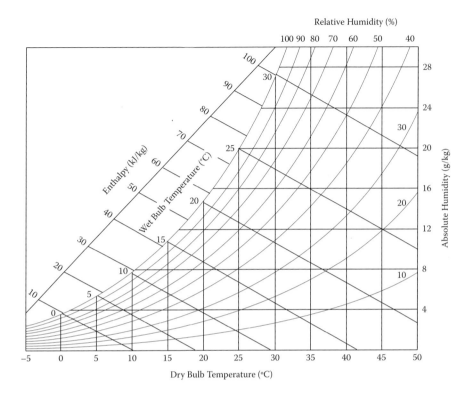

FIGURE 3.2 Simplified psychrometric chart.

expand upward as the temperature increases toward the right of the chart because the air can hold more water at higher temperatures. The top curve is the saturation line, which indicates the maximum amount of moisture that the air can hold at that specific temperature. If the dry-bulb temperature drops below the saturation line, then condensation occurs. The amount of this condensation can be read vertically on the moisture content scale.

Enthalpy is the heat content of the air relative to 0°C and 0% relative humidity. It is the sum of the latent and sensible heat and is measured in kJ/kg or Btu/lb dry air. Enthalpy has two components, sensible heat and latent heat. Sensible heat increases the dry-bulb temperature of the air at a rate of about 1.005 kJ/kgK, and latent heat is the heat that is necessary to evaporate liquid water to form the moisture content of the air. The enthalpy scale is usually indicated outside the body of the chart toward the upper left (Figure 3.6). The wet-bulb temperature lines are almost parallel to the enthalpy lines.

Several psychrometric processes can be plotted on the chart. For designers the most important of these processes are heating, cooling, humidification, and evaporative cooling.

Heating the air will move a point horizontally to the right (A to C), and cooling the air will move it horizontally to the left (A to B) (Figure 3.7). There is no change in absolute humidity; however, the relative humidity changes because of the differences

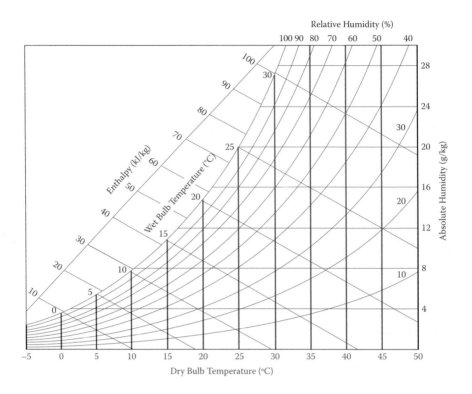

FIGURE 3.3 Dry-bulb temperature in the psychrometric chart.

in the maximum amount of moisture that the air can hold at different temperatures. If the temperature is cooled to the saturation line, then condensation will occur and the point at which this occurs is the dew point temperature.

When water particles evaporate within a stream of ambient air without a supply of external heat, the temperature of the air is lowered and its moisture content is elevated, while its wet-bulb temperature remains constant (A to B). This change in air temperature, which occurs without addition or extraction of heat from the system, following the enthalpy lines, is called "adiabatic," meaning that the total energy content of the air (sensible plus latent) is kept constant (Figure 3.8). When air is passed through a chemical sorbent such as silica gel, this material removes some of the moisture content, reducing the absolute and relative humidity and increasing its heat content because of the release of heat through this process (A to C).

3.2 THERMAL COMFORT

The ASHRAE definition of comfort is "that condition of mind that expresses satisfaction with the thermal environment" (ASHRAE, 2005). In general, comfort occurs when body temperatures are held within specific narrow ranges, skin moisture is low, and the physiological effort of regulation is minimized. We are comfortable when we don't have to think about it.

Thermal Comfort

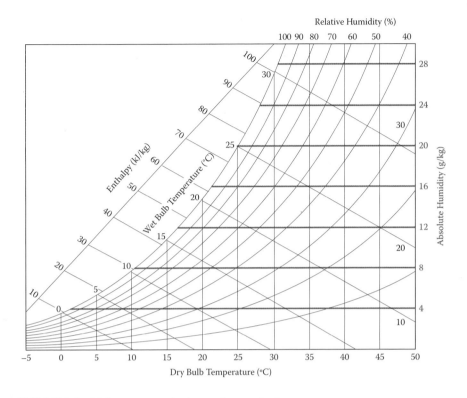

FIGURE 3.4 Moisture content in the psychrometric chart.

The human being burns food as fuel and generates heat. About 20% of the energy produced during this biological process is turned into work and the rest is dissipated as waste heat. If the body is not able to dissipate this heat it will overheat. Heat-flow mechanisms (conduction, radiation, and convection) dissipate heat and maintain an optimum operating temperature in our body.

3.2.1 Heat Balance

To operate adequately, the human body must be in heat balance. Warm-blooded animals require a very constant temperature to operate, and our body tries to maintain a deep body temperature of about 37°C (98.6°F) and a skin temperature of about 33°C (91.4°F). If the human body heats up too much then it will not function properly. At about 38°C (100°F) our body begins to sweat, at about 40°C (104°F) we have a nauseous feeling, at about 42°C (107.6°F) we pass out, and if our body temperature reaches 44°C (111°F) we will have brain damage. If our body cools down too much then it will not function properly either. At about 35°C (95°F) we start shivering and have goose bumps, at about 29°C (84.2°F) we lose the power of speech, at about 22°C (71.6°F) the body is too cold for most situations, and at about 15°C (59°F) it is virtually impossible to recover.

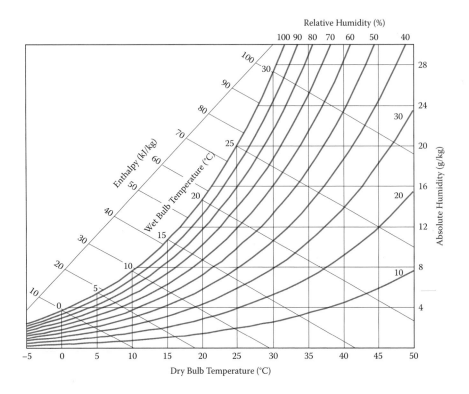

FIGURE 3.5 Relative humidity in the psychrometric chart.

To maintain optimum operating conditions and avoid overheating or overcooling, the heat that is generated by the metabolism or gained from the environment must be dissipated at the same rate that it is gained. The body is homoeothermic, which means that it has several physiological regulatory mechanisms to control heat dissipation. When the air is warm, the body responds by vasodilation: Subcutaneous blood vessels expand and increase the skin blood supply and the skin temperature, which in turn increases heat dissipation. If this is not enough, then the body will begin to sweat (evaporative cooling) using body heat to evaporate sweat. If this is not sufficient to restore balance, then body heating or hyperthermia, which is insufficient heat loss, will occur. When the air is cold, the response is vasoconstriction, reducing blood circulation to the skin and thus lowering the skin temperature and the rate of heat dissipation. Pores are also closed, reducing surface evaporation. Shivering also increases the production of metabolic heat. If these mechanisms are not sufficient to reduce heat losses, then the body will cool and hypothermia (i.e., excessive heat loss) will occur.

The sensation of comfort depends a great deal on the ease with which the body is able to regulate the balance. The body will always be producing energy, and there will always be heat gains and losses; however, the body temperature must remain stable. The balance between heat gains and losses is expressed in (3.1).

$$\Delta S = M \pm R_d \pm C_v \pm C_d - E_v \tag{3.1}$$

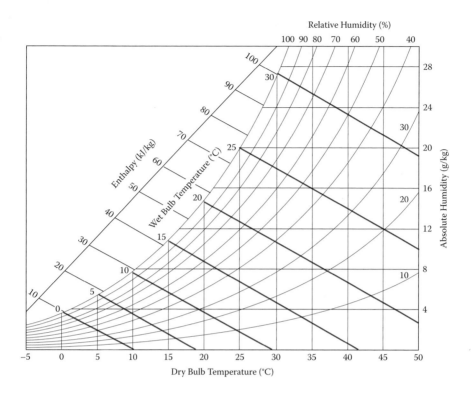

FIGURE 3.6 Enthalpy and wet-bulb temperature in the psychrometric chart.

where
  ΔS = change in stored heat
  M = metabolic heat production
  R_d = net radiation exchange
  C_v = convection
  C_d = conduction
  E_v = evaporation
  ΔS = change in stored heat

Evaporation is always negative (heat dissipation mechanisms), metabolic heat production is always positive (heating mechanisms) and the rest can be positive or negative depending on the direction of the heat flow. Figure 3.9 shows an example in which the temperature of the body is higher than the temperature of the floor so the heat is flowing from the body to the floor (negative); the temperature of the wall is lower than the temperature of the body, so the heat is flowing from the body to the wall (negative); the temperature of the air is warmer than the temperature of the body (positive heat flow); and the net radiant exchange between the body and the sun is positive. For the body to be in equilibrium, ΔS must be zero. This is a precondition, but not the only variable that determines thermal comfort.

The human body maintains a balance with its environment and must dissipate the heat that is generated or absorbed by it. At about 21°C (70°F), the body heat

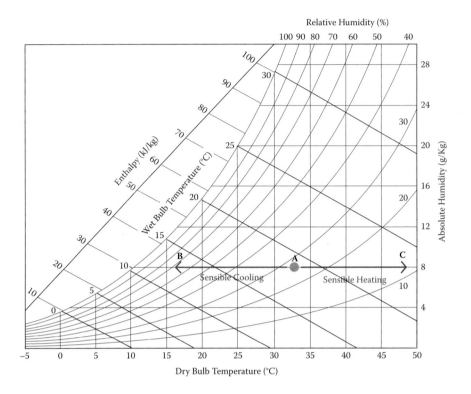

FIGURE 3.7 Sensible heating and sensible cooling in the psychrometric chart.

losses are 40% by convection, 20% by evaporation, 40% by radiation, and just a little by conduction. However, these ratios are affected by air temperature; as the air temperature increases, the amount of heat lost by evaporation increases, and above a temperature of 37°C (98.6°F) sweating must dissipate all the heat that is generated by the body plus the heat that is added by radiation, convection, and conduction (Figure 3.10).

3.2.2 Variables That Affect Thermal Comfort

Thermal comfort is affected by several factors that affect the rate of heat dissipation from the body and are usually classified as environmental or personal factors. There are also additional contributing factors that affect thermal comfort.

The environmental factors are air temperature, radiation, air motion, and relative humidity. Air temperature measured by the dry bulb temperature (DBT) is probably the single most important environmental factor, because together with air movement it affects the rate of convective heat dissipation. The air temperature is measured using a dry-bulb thermometer.

Radiation affects thermal comfort through the radiant exchanges between the body and the surfaces that surround it. The magnitude of this exchange is affected by the difference in temperature between our body and the exchange surface and the view angle between this surface and our body. A larger angle or a higher

Thermal Comfort

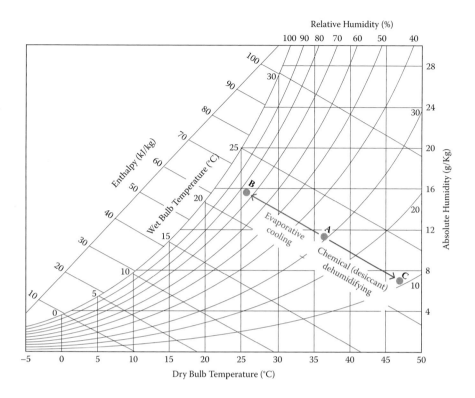

FIGURE 3.8 Adiabatic cooling and heating.

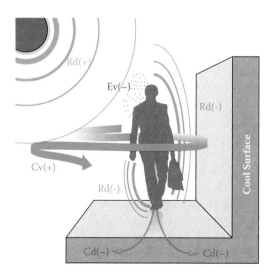

FIGURE 3.9 Heat transfer mechanisms.

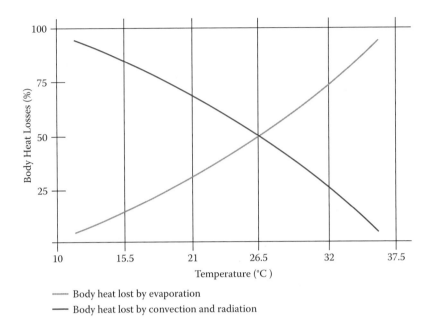

FIGURE 3.10 Heat dissipation for a person at rest at different temperatures.

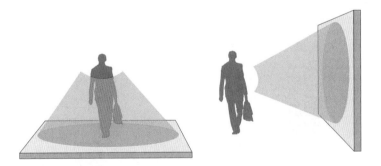

FIGURE 3.11 Larger angle of exposure increases radiant exchange.

temperature difference between the two surfaces will produce a larger radiant exchange (Figures 3.11 and 3.12) with new radiation always moving from a higher to a lower temperature. The thermal effect of the surfaces surrounding and affecting an object is indicated by the mean radiant temperature (MRT). This is measured with a globe thermometer, which is usually a matte black copper sphere 15 cm in diameter that responds to the effect of radiation flows from the surrounding surfaces. If the temperature of the globe thermometer is higher than the DBT, then the surrounding surfaces are warmer than the air; if it is lower, then the surface temperatures are lower than the temperature of the air.

Air motion affects the evaporation of moisture from the skin. A higher air velocity increases the evaporative cooling effect and reduces the surface resistance of the body. Air movement is one of the most useful strategies to cool the body in a hot and

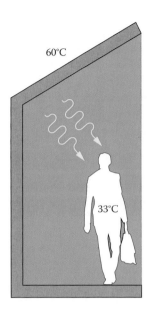 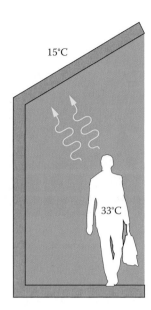

FIGURE 3.12 Higher temperature difference increases radiant exchange.

humid climate and is measured in meters per second (m/s) or feet per minute (ft/min). Higher air velocities, up to about 2 m/s, can increase the sensation of thermal comfort in overheated conditions.

Relative humidity affects the evaporation rate from the skin. A lower relative humidity of the air allows more sweat to evaporate and cool the skin, whereas higher relative humidity (i.e., above 70%) makes it more difficult for the sweat to evaporate because the air's moisture content is already elevated. In humid climates, sweat will not evaporate because the air is already close to the saturation level. Very low relative humidity (i.e., below 30%) is also uncomfortable because of its drying effect on the mucous membranes. Humidity between 30 and 60% is usually considered acceptable because it has a limited effect on thermal comfort.

The personal variables that affect thermal comfort are the metabolic rate (activity) and clothing. The metabolic rate is the heat generated by the body. Heat production in the human body can be of two kinds: (1) basal metabolism due to biological processes that are continuous and nonconscious and are in the range of about 60 W/m^2 of body surface area (about 100 watts in a typical body); and (2) muscular metabolism, which is produced while carrying out work that is consciously controllable (except shivering). Metabolic rate can be expressed as power density in mets or W/m^2, where 1 met = 58.2 W/m^2 of body surface area (18.4 Btuh/ft^2 of body surface area) and varies depending on the activity level of the body (Figure 3.13, Table 3.1).

Clothing is insulation around the body, which means that it can be measured in the same units of thermal transmittance (U-value) W/m^{2}K. To simplify the units when studying thermal comfort, a unit called the clo has been created. One clo has the U value of 6.45 W/m^{2}K, which is equal to a resistance of 0.155 m^{2}K/W spread over the whole body. It is probably easier to visualize by understanding that 1 clo is equivalent to the insulating value of trousers, T-shirt, long-sleeve

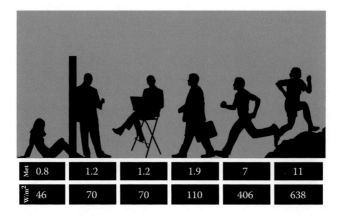

FIGURE 3.13 Activity levels and metabolic rates.

shirt, plus long-sleeve sweater. Table 3.2 lists clo values for several typical clothing ensembles. Additional tables in the *Handbook of Fundamentals* (ASHRAE, 2009) describe the insulation values of individual ensembles, which can be used to calculate the clo value with more precision.

In addition to personal and environmental factors, other contributing factors mostly affect the metabolic rate. Food and drink influence the amount of energy that must be dissipated. Acclimatization includes short-term physiological and long-term endocrine adjustments to climate. Body form and mass affect the metabolic rate. The heat production is proportional to the amount of body mass; more mass produces more heat and the heat dissipation is proportional to the body surface area. Thus, a person with a higher surface-to-volume ratio can dissipate more heat and has more tolerance to higher temperatures. Subcutaneous fat is an insulating layer. The elderly have narrower comfort ranges because their adjustment mechanisms are not working as efficiently, making it more important to provide them with an appropriate comfort zone.

Some authors now include psychological factors such as color, texture, sound, light, movement, and aroma; however, it is difficult to find scientific evidence that demonstrates the effect of these variables.

3.3 ENVIRONMENTAL AND COMFORT INDICES

Environmental and comfort indices simplify the description of the thermal environment and the stress that it imposes on the human body because they combine two or more parameters (e.g., air temperature, mean radiant temperature, humidity, air velocity) into a single variable. Many environmental and comfort indices have been developed over the years, and their discussion is outside the scope of this book. Only two will be discussed in this section.

The operative temperature is a simple and easy-to-use environmental index that combines into a single number the effects of dry-bulb temperature, radiant temperature, and air motion on the sensation of warmth or cold felt by the human body. Without considering air velocities, the simplest, though not perfectly accurate, way

TABLE 3.1
Rates of Metabolic Heat

Activity	Btu/h-ft^2	W/m^2	Met*
Resting			
Sleeping	13	40	0.7
Reclining	15	45	0.8
Seated, quiet	18	60	1.0
Standing, relaxed	22	70	1.2
Walking (on level surface)			
2.9 fps (2 mph)	37	115	2.0
4.4 fps (3 mph)	48	150	2.6
5.9 fps (4 mph)	70	220	3.8
Office Activities			
Seated, reading or writing	18	60	1.0
Typing	20	65	1.1
Filing, seated	22	70	1.2
Filing, standing	26	80	1.4
Walking about	31	100	1.7
Lifting/packing	39	120	2.1
Driving/Flying			
Car	18–37	60–115	1.0–2.0
Aircraft, routine	22	70	1.2
Aircraft, instrument landing	33	105	1.8
Aircraft, combat	44	140	2.4
Heavy vehicle	59	185	3.2
Miscellaneous Occupational Activities			
Cooking	29–37	95–115	1.6–2.0
House cleaning	37–63	115–200	2.0–3.4
Seated, heavy limb movement	41	130	2.2
Machine work			
sawing (table saw)	33	105	1.8
light (electrical industry)	37–44	115–140	2.0–2.4
heavy	74	235	4.0
Handling 110 lb bags	74	235	4.0
Pick and shovel work	74–88	235–280	4.0–4.8
Miscellaneous Leisure Activities			
Dancing, social	44–81	140–255	2.4–4.4
Calisthenics/exercise	55–74	175–235	3.0–4.0
Tennis, singles	66–74	210–270	3.6–4.0
Basketball	90–140	290–440	5.0–7.6
Wrestling, competitive	130–130	410–505	7.0–8.7

Source: ASHRAE Standard 55-2004 *Thermal Environmental Conditions for Human Occupancy,* 2004.

Note: MET: metabolic heat rate.

TABLE 3.2
Clothing Insulation Values for Typical Ensembles

Clothing Description	Garments Included[a]	(clo)
Trousers	1. Trousers, short-sleeve shirt	0.57
	2. Trousers, long-sleeve shirt	0.61
	3. #2 plus suit jacket	0.96
	4. #2 plus suit jacket, vest, T-shirt	1.14
	5. #2 plus long-sleeve sweater, T-shirt	1.01
	6. #5 plus suit jacket, long underwear bottoms	1.30
Skirts and dresses	7. Knee-length skirt, short-sleeve shirt (sandals)	0.54
	8. Knee-length skirt, long-sleeve shirt, full slip	0.67
	9. Knee-length skirt, long-sleeve shirt, half slip, long-sleeve sweater	1.10
	10. Knee-length skirt, long-sleeve shirt, half-slip, suit jacket	1.04
	11. Ankle-length skirt, long-sleeve shirt, suit jacket	1.10
Shorts	12. Walking shorts, short-sleeve shirt	0.36
Overalls and coveralls	13. Long-sleeve coveralls, T-shirt	0.72
	14. Overalls, long-sleeve shirt, T-shirt	0.89
	15. Insulated coveralls, long-sleeve thermal underwear tops and bottoms	1.37
Athletic	16. Sweat pants, long-sleeve sweatshirt	0.74
Sleepwear	17. Long-sleeve pajama tops, long pajama trousers, short three-quarter length robe (slippers, no socks)	0.96

Source: ASHRAE Standard 55-2004 for Thermal Environmental Conditions for Human Occupancy, 2004.

[a] All clothing ensembles, except where otherwise indicated in parentheses, include shoes, socks, and underwear. All skirt/dress clothing ensembles include pantyhose and no additional socks.

to calculate operative temperature at low air velocities is to average the dry-bulb temperature and the mean radiant temperature (Equation (3.2)). The relationship indicated by Equation (3.2) is affected by the clothing level. If the subject is dressed lightly as is customary in a hot climate, the radiant effect is higher than the effect of the air temperature.

$$OT = (DBT + MRT)/2 \qquad (3.2)$$

where
 OT = operative temperature
 DBT = dry-bulb temperature
 MRT = mean radiant temperature

Other equations combine several environmental variables to define a comfort scale. The most important of these is the predicted mean vote (PMV), which predicts the comfort sensation and the degree of discomfort (thermal dissatisfaction) of people exposed to moderate thermal environments. PMV is discussed in the next section.

TABLE 3.3
Equations for Predicting Thermal Sensation (Y) of Men, Women, and Men and Women Combined

Exposure Period in Hours	Sex	Regression Equations t = dry-bulb temperature, °C p = vapor pressure, kPa
1.0	Male	$Y = 0.220t + 0.233p - 5.673$
	Female	$Y = 0.272t + 0.248p - 7.245$
	Combined	$Y = 0.245t + 0.248p - 6.475$
2.0	Male	$Y = 0.221t + 0.270p - 6.024$
	Female	$Y = 0.283t + 0.210p - 7.694$
	Combined	$Y = 0.252t + 0.240p - 6.859$
3.0	Male	$Y = 0.212t + 0.293p - 5.949$
	Female	$Y = 0.275t + 0.255p - 8.622$
	Combined	$Y = 0.243t + 0.278p - 6.802$

Source: Adapted from 2005 *ASHRAE Handbook of Fundamentals* (with permission).

Another set of equations to predict thermal sensation was developed by Rohles (1973) and Rohles and Nevins (1971) and used by ASHRAE (2005) to predict thermal sensation as a function of the exposure period, dry-bulb temperature, and vapor pressure. The author developed different equations for men and women to predict the thermal comfort sensation (Table 3.3). The ideal value for the comfort sensation (Y) is zero, and in these equations there are several assumptions: the activity level is sedentary, clothing is 0.5 clo, $MRT = DBT$, and air velocity is below 0.2 m/s.

3.4 COMFORT MODELS

The two main theories of comfort are the heat balance or physiological approach and the adaptive approach. In simple terms, the physiological or heat balance model states that humans are all the same and that comfort on all subjects can be explained by a physiological approach. In this model the comfort zone is the same for all without considering location and adaptation. On the other hand, the adaptive model states that in occupant-controlled, naturally conditioned spaces, the thermal response and thus the comfort zone are affected by outdoor climate.

3.4.1 PHYSIOLOGICAL COMFORT MODEL

The physiological comfort model is best exemplified by the PMV/PPD model, in which PPD is the predicted percent of dissatisfied people at each PMV; the model uses the Fanger PMV formula. PMV establishes a thermal strain based on steady-state heat transfer between the body and the environment and assigns a comfort vote to that amount of strain.

The original data to develop Fanger's PMV model were collected by subjecting a large number of people to different conditions within a climate chamber and having them select a position on the scale that best described their comfort sensation. A mathematical model of the relationship among all the environmental and physiological factors considered was then derived from the data. The PMV is the average comfort vote predicted by a theoretical index for a group of subjects when subjected to a particular set of environmental conditions. The PMV uses a seven-point thermal scale that runs from cold (–3) to hot (+3) with zero as the ideal. Lower or higher values are possible. From the PMV, the PPD can be determined. As PMV moves away from zero in either the positive or negative direction, PPD increases. The maximum number of people that could be dissatisfied with their comfort conditions is 100%. However, as you can never please everybody even under optimum conditions, the minimum PPD number even under optimum comfort conditions is 5%. Fanger's method is prevalent in most thermal comfort standards such as ISO 7730. This standard predicts thermal comfort comparing four physical variables, air temperature, air velocity, mean radiant temperature, and relative humidity with two personal variables—clothing insulation and activity level (Fanger, 1967).

A commonly used method to determine if thermal conditions are comfortable is to plot them in a psychrometric chart with the comfort zone plotted in it. An example is ASHRAE's comfort zone for typical indoor environments, which uses operative temperature and humidity ratio (Figure 3.14). ASHRAE standard 55-2004 states that 80% of occupants would find these conditions acceptable. This value is based on 10% dissatisfaction criteria for whole-body thermal comfort that uses the PMV–PPD index plus an additional 10% dissatisfaction that may occur from partial body thermal discomfort. Two comfort zones are indicated in this figure, one for 0.5 clo (summer clothing) and one for 1.0 clo (indoor winter clothing). These zones are valid when air speeds are below 0.20 m/s, and the activity level of the occupants is moderate or low with a metabolic rate between 1.0 met and 1.3 met.

3.4.2 Adaptive Comfort Model

Researchers doing field investigations with subjects engaged in typical daily activities in their environments, instead of experimental chambers, have found that the subjects' responses in familiar daily situations were different than the predictions using the PMV model, especially when the building was not mechanically heated or cooled. Even though these field studies do not have the scientific rigor of laboratory experiments, they are more immediately relevant to typical living or working conditions. Some authors (Fanger and Toftum, 2002) have tried to explain these differences as differences in expectations in terms of their exposure level to buildings that are mechanically heated and cooled.

Thermal neutrality temperature (T_n) is the air temperature at which a large group of people would not feel hot or cold and the body would be under minimal thermal stress. Probably the first review of available field data that found a statistical relationship between mean levels of dry-bulb temperature and thermal neutrality

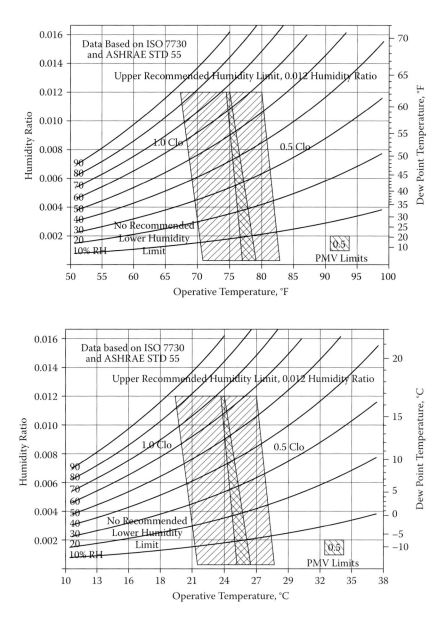

FIGURE 3.14 Acceptable range of operative temperature and humidity for occupants that have metabolic rates between 1.0 and 1.3 met, clothing is between 0.5 clo and 1.0 clo, air speeds are under 0.2 m/s. (From *ASHRAE Standard 55-2004*.)

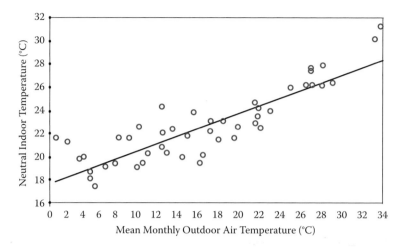

FIGURE 3.15 Auliciems's correlation of indoor neutrality with monthly mean outdoor temperature. (From Auliciems, A. and Szokolay, S. PLEA Notes 3: Thermal Comfort 1997 in association with the University of Queensland Department of Architecture.)

temperature was done by Humpreys as far back as 1975. In 1981, Auliciems used a larger database and also found a correlation between the thermal neutrality in free-running buildings and the outdoor average temperature (Figure 3.15). This correlation was valid for temperatures between 18°C (64.4°F) and 28°C (82.4°F) (Auliciems and Szokolay, 1997) expressed in the following equation:

$$T_n = 17.6 + 0.31\ T_{ave} \qquad (3.3)$$

where
T_n = thermal neutrality temperature
T_{ave} = outdoor average dry-bulb temperature

Griffiths (1990) and Nicol and Roaf (1996) found similar correlations between the neutrality temperature and the outdoor average temperature. In general these equations are valid when the outdoor temperature is between 17°C (62.6°F) and 30°C (86°F). The comfort zone lies between 2°C above and below the neutral temperature. De Dear and Brager (1998) examined the relationship between thermal comfort and indoor and outdoor temperature in a database of 21,000 observations from 160 buildings worldwide. They found that the PMV model fully explained adaptation in buildings with heating, ventilating, and air-conditioning (HVAC) systems, but occupants in naturally ventilated buildings were tolerant of a significantly wider range of temperatures. They explained this by a combination of both behavioral adjustments and physiological adaptations. The results of this research were incorporated in a revision to ASHRAE standard 55. This was implemented in the 2000 revision of ASHRAE comfort standard 95 and is valid only when mean the monthly outdoor temperature is between 10°C (50°F) and 33.5°C (92.3°F) (Figure 3.16). This standard, which was originally based solely on the PMV model, now has an optional

Thermal Comfort

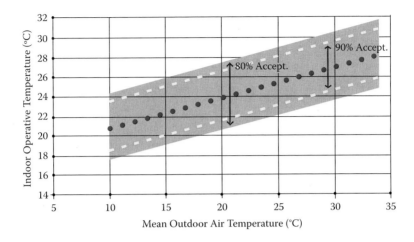

FIGURE 3.16 Acceptable operative temperature ranges for naturally conditioned spaces. (From *ASHRAE Standard 55-2004 for Thermal Environmental Conditions for Human Occupancy*, 2004.)

method for determining acceptable thermal conditions in naturally conditioned spaces. The *ASHRAE Standard 55* (2010) states that occupants' thermal responses in naturally conditioned spaces depend in part on the outdoor climate and may differ from thermal responses in buildings with centralized air conditioning systems. This is due to differences in recent thermal experiences, changes in clothing, availability of control options, and shifts in occupant expectations.

The method applies only to spaces without air conditioning and with operable windows that open to the outdoors. Mechanical ventilation with unconditioned air may be used, but opening and closing of windows must be the primary means of regulating conditions in the space. The space may have a heating system, but this optional method does not apply when the heating system is in operation. The occupants must be engaged in near-sedentary activities with metabolic rates from 1.0 to 1.3 met. The users of a space can undertake three types of adaptive processes to achieve thermal comfort: (1) behavioral (e.g., clothing, body movements or adjustments such as windows to modify the body's heat balance); (2) physiological (e.g., vasodilation, vasoconstriction, shivering, sweating); and (3) psychological adaptations (e.g., thermal perceptions and expectations). Because of these adaptations, neither humidity nor air speed limits need to be considered in this model. However, with high humidity (i.e., above 70%) it is difficult to believe that temperatures in the upper portion of the chart (i.e., above 29°C) will be neutral temperatures.

The neutrality temperature in this model can be calculated using the following correlation:

$$T_n = 18.9 + 0.255\, ET_{out} \tag{3.4}$$

where
ET_{out} = outdoor average effective temperature that accounts for humidity
T_n = neutrality temperature

The implementation of the adaptive model in the ASHRAE comfort standard is important because it accepts that naturally ventilated buildings, in which the occupants have more control over their thermal environment, have a larger comfort zone. A larger comfort zone in naturally ventilated buildings promotes the implementation of passive cooling strategies in buildings, highlighting the role of the user in controlling the thermal environment.

ASHRAE's standard 55-2004 provides an option for adaptive comfort derived from neutralities and temperature preferences in naturally ventilated spaces as a function of prevailing outdoor temperature. In this standard there is a chart indicating the temperature ranges that correspond with thermal acceptability ratings of 90% and 80% (Figure 3.16). The value of $T_{a(out)}$ is the average outdoor monthly air temperature calculated using the arithmetic average of the mean monthly minimum and maximum daily air temperatures for the month in question. The preferred temperature inside a naturally ventilated building increases by approximately 1 degree for every 3 degree increase in mean monthly outdoor air temperature (Brager and de Dear, 2000). The linear equation for the optimum comfort model described by Figure 3.16 and its acceptability limits, collectively known as the Adaptive Comfort Standard (ACS), can be written as

$$T_n = 0.31 \times T_{ave} + 17.5 \qquad (3.5)$$

where

T_{ave} = monthly mean outdoor air temperature
T_n = neutrality temperature

The European standard for naturally ventilated buildings (EN15251) now also incorporates a version of an adaptive method developed by Nicol and Humphreys (2010). The equation for thermal comfort for buildings in the free-running mode (Annex A2) in European Standard EN15251 rests on the data collected in the EU project Smart Controls and Thermal Comfort (SCATs). The equation proposed by Nicol and Humphreys is

$$T_n = 0.33 \times T_{ave} + 18.8 \qquad (3.6)$$

where

T_{ave} = monthly mean outdoor air temperature
T_n = neutrality temperature

The equation and charts proposed for this standard are similar, but not identical, to those proposed by the ASHRAE standard. The differences between the two equations are probably a result of several differences in the research. The databases are different because ASHRAE 55-2004 uses the data from the ASHRAE world database of field experiments collected by de Dear and Brager (1998), while EN1521 uses the data from the more recent European SCATs project. The ASHRAE adaptive standard applies only to naturally ventilated buildings, whereas the EN15251 applies to any building in free-running mode, thus including more buildings. The derivation of the neutral temperature is different, and the outdoor temperature is defined differently; ASHRAE is defined in terms of the monthly mean outdoor temperature, while EN15251 uses a running average.

It is now generally accepted that people in free-running buildings may also make various adjustments to themselves and their surroundings to reduce physiological strain. This adaptive approach is a behavioral approach because it assumes that people will respond to variations by adjusting to them through conscious actions related to their bodies or their environments. Bodily adjustments are primarily connected to clothing, posture, activity schedules, and activity levels, and environmental adjustments are connected to ventilation, air movement, and local temperature. They might also include unconscious long-term changes to physiological set points for control of shivering, skin blood flow, and sweating.

3.5 THE PERCEPTION OF COMFORT

Traditional comfort standards aim to achieve a thermal "steady state" across time and a thermal equilibrium across space (Fitch, 1947). There is an underlying assumption that the best thermal environment never needs to be noticed and that environmental control systems are directed to produce standard comfort conditions. However, this is not necessarily true; changes in the perception of the thermal environment help us connect with the daily rhythms of the environment that surrounds us. Furthermore, as climate becomes more erratic and extreme, it will become more challenging to achieve stable comfortable environments without substantially increasing energy use and emissions.

It is a fact that thermal comfort has a physiological basis and is affected by the heat balance of the human body and the four environmental and personal variables discussed in this chapter. However, comfort zones should not be static and narrow. Thermal comfort is affected by other variables. The temperature that most people find comfortable in naturally ventilated buildings, and maybe in all passive buildings, will vary with the outdoor temperature as people change their behavior to adapt to the changes in outdoor temperature, sometimes even anticipating it, reinforcing our connection with the environment that surrounds us. This process establishes a strong connection with the outdoor environment and nature's rhythms. Furthermore, the nature of this adaptive process is very complex, and other factors probably have not been quantified yet and also affect thermal comfort. Psychological, social, and cultural factors most likely have more influence than is currently accepted. For example, in the same developing country, people who live, work, and travel in air-conditioned spaces will probably have a different comfort zone from others who live, work, and travel in unconditioned spaces.

The fact that adaptive comfort is gaining importance and is being incorporated into codes is significant because buildings with wider comfort bands have better opportunities to implement passive heating or cooling systems. The implementation of adaptive comfort as part of a thermal comfort standard is a step in the right direction. However, more should be done. Kwok and Rajkovich (2010) suggest a mesocomfort zone that exists between the "optimum" thermal comfort conditions specified by static models and thermal environmental conditions that produce involuntary responses and that would provide opportunities for adaptation. The size of this zone would be expected to be proportional to the adaptive capacity of a space when combined with the opportunities afforded to the occupant. Some of these opportunities are directly linked to the building design.

If the comfort zone becomes a sliding band that moves with the seasons and building users have a more active role in maintaining comfort inside a building, it will be easier to design comfortable low-carbon buildings than if fixed, narrow thermal comfort bands with no user control are used. If we understand that users have more potential to participate in the control of their environment and that a building can provide thermal controls through other means than the simply mechanical, we will be closer to the creation of low-carbon, climate-responsive, living buildings and farther from the sealed, inefficient, carbon-emitting buildings of today.

4 Climate and Architecture

4.1 CLIMATE

Differences in climate are caused by the differential solar heat input and the almost uniform heat emission over the Earth's surface. Equatorial regions receive much greater energy than areas near the pole, and this differential, together with, for example, the rotation of the Earth and the presence of land or water masses and mountain ranges, is the main driving force of atmospheric phenomena.

Weather is generally assumed to be the set of atmospheric conditions prevailing at a given place and time, and *climate*, from the ancient word *clime*, is usually thought of as the integration in time of weather conditions, characteristic of a certain geographical location (Docherty and Szokolay, 1999).

4.2 CLIMATE AND ARCHITECTURE

The amount of energy used in a building is a direct result of the interactions among the external environment, the building's program, and the building's form. Buildings should be able to regulate outdoor conditions and to harness available energy to achieve internal comfort with minimum emissions.

The climate where the building is located can be understood through the analysis of data, which are collected in meteorological stations and are published in summary form usually. Examples of these data include the following:

- Dry-bulb temperature
- Relative and absolute humidity
- Air movement, measured as wind speed and direction
- Precipitation, measured as the total amount of rain, hail, snow, or dew in mm per unit time (e.g., day, month, year)
- Cloud cover based on visual observation and expressed as a fraction of the sky hemisphere covered by clouds (tenths, or octas = eighths)
- Sunshine duration, the period of clear sunshine (when a sharp shadow is cast), measured by a sunshine recorder that burns a trace on a paper strip and expressed as hours per day or month
- Solar radiation, measured by a pyranometer on an unobstructed horizontal surface, usually recorded as the continuously varying irradiance (W/m^2)

The most important climate variables in building design are those that directly affect thermal comfort: temperature, humidity, solar radiation, and air movement. Precipitation data are helpful in designing drainage systems and slopes; even though

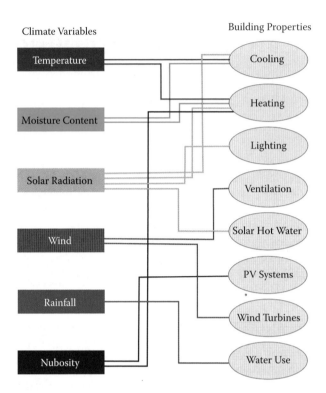

FIGURE 4.1 Effects of climate variables on building design and emissions.

these data do not directly affect building energy use, they are correlated with other climate parameters that affect building energy use: solar radiation, cloud cover, relative humidity, diurnal temperature swing, and wet- and dry-bulb design temperatures.

Furthermore, climate variables also affect the building design and emissions in different ways. Temperature affects heating and cooling loads; moisture content of the air also affects cooling loads and in some degree heating loads; radiation affects heating, cooling, daylighting, and active systems such as domestic solar hot water and photovoltaic systems; wind affects airflow inside buildings and renewable systems such as windmills; precipitation affects water use; cloud cover affects heating, cooling, daylighting, and photovoltaics. Figure 4.1 shows the relationship between climate variables and some building properties.

To study year-round building performance, annual weather data are required. A minimum amount of data is necessary for an adequate design:

1. Temperature: monthly mean of daily maxima and daily minima (°C or °F)
2. Humidity: minimum and maximum mean relative humidity (%)
3. Solar radiation: monthly mean daily total (in MJ/m^2 or Wh/m^2)
4. Sunshine: (%)
5. Wind: prevailing wind speed (m/s or mph) and direction
6. Rainfall: monthly total (in mm or inches)

For an easier understanding of the data, the raw weather data from the meteorological stations are usually presented in tabular or graph form. Some design handbooks and standards such as American Society of Heating, Refrigerating and Air-Conditioning Engineers (ASHRAE) also provide general climatic data for building design and manual load calculations.

The degree day is a unit of measurement that indicates the heating or cooling demand in a building and helps to define the severity of the climate. There are two types of degree days: heating degree days (HDD) and cooling degree days (CDD). A HDD indicates the energy required to heat the building and is defined relative to a base temperature, which is the outside temperature above which the building does not need any heating. This base temperature is affected by building design characteristics such as the level of insulation and size and position of the openings, but also by other factors such as the amount of solar radiation falling on the building or the amount of heat generated by people or appliances inside the building. A tighter well-insulated building will have a lower base temperature than a leaky noninsulated building. In the United States HDD tables have usually been provided with base temperatures of 18.3°C (65°F) or 15.5°C (60°F). A HDD can be calculated at any time interval, and shorter intervals provide more precision. However, it is usually calculated using average daily values which are then added to determine the yearly total. For example if the reference temperature is 18.3°C (65°F) and the average outdoor temperature for a given day is 10°C (15°F), then the HDD for that day is 8.3°C F. This would be done for every day of the year to determine the yearly total. A CDD indicates the amount of energy required to cool the building and is defined relative to a base temperature, which is the outside temperature above which the building needs to be cooled. CDD are usually calculated from the same base temperature as HDD, thus whenever the outside temperature is above this base temperature cooling will be required and whenever it is below, heating will be required. With the advent of energy simulation tools and more precise methods to determine energy use in a building, the use of HDD and CDD is declining; however, it provides an easy way to visualize the severity of a winter or a climate in a given location.

The development and generalized use of detailed building energy simulation programs has generated the need for hourly data for whole years. In a short time, hourly data have become the standard for climate analysis and energy simulation. Most data set systems construct a composite year's data by selecting periods from actual data over many years of recording. These types of data have different names depending on the country: In the United States, typical meteorological year (TMY) and EnergyPlus weather files (EPW); in the United Kingdom, the Chartered Institution of Building Services Engineers (CIBSE); in China, the Chinese Standard Weather Data (CSWD); in Australia, the representative meteorological year (RMY); in New Zealand, the New Zealand Data (NiWA); and in Spain, the Spanish Weather for Energy Calculations (SWEC). These files are powerful because they provide hourly data for the whole year; however, because of the process by which they are generated, the values in the files might not accurately reflect conditions in a given location.

Precipitation, terrain, vegetation, degree of solar exposure, wind patterns, the presence of water bodies, geology, and the influences of buildings or other built forms on

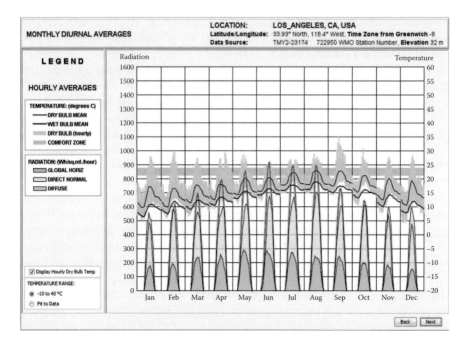

FIGURE 4.2 Monthly climate data plotted with Climate Consultant.

or near the site can create unique, site-specific, climatic conditions (Crowther, 1984) that could be very different from those of the closest weather station and thus can considerably affect the building and its design. The architect should try to determine microclimate effects by collecting data on site through low-cost weather stations or by trying to determine the microclimatic effects with available data.

Climate graphs and charts can now be created from yearly data using climate analysis software. Figures 4.2 and 4.3 show climate data generated with two widely used climate analysis tools: Climate Consultant, developed at the University of California, Los Angeles (UCLA) by Murray Milne and Robin Liggett; and the Weather Tool, now part of Ecotect. These charts select appropriate information from the weather files and make it possible to understand the otherwise illegible yearly weather files. Figure 4.2 shows monthly diurnal averages of climate data for Los Angeles generated with Climate Consultant. It includes monthly mean and hourly values of dry-bulb temperature, mean wet-bulb temperature, global direct radiation, and diffuse solar radiation. The Weather Tool inside Ecotect can also show data in this form and can generate a weekly summary of the different variables. Figure 4.3 illustrates the weekly summary for maximum temperature using the same weather file for Los Angeles during the 24 hours and 52 weeks of the year.

4.3 CLIMATE ZONES

There are many climate classification systems, most of them based on vegetation types or amounts of precipitation. In 1884 Köppen proposed a climate

Climate and Architecture 101

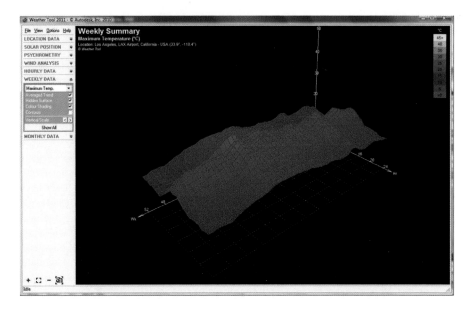

FIGURE 4.3 Weekly temperature data plotted with Ecotect's Weather Tool.

classification system based on the concept that native vegetation provides the best way to define climate types. Köppen (1918) revised this classification system several times including one with Geiger, which is why it is sometimes called the Köppen–Geiger system. This system is still in use and consists of several climate groups subdivided into climate types and subtypes. Climate zone boundaries are selected according to the vegetation distribution, and then includes quantitative definitions based on temperature and precipitation indices with their seasonality distribution. This classification scheme divides the climates into five main groups and several types and subtypes, and uses two- and three-letter symbols to designate the climate types.

Even though Köppen's climate classification system is the most widely used all over the world, it is not very useful for architectural purposes because it does not directly provide the information necessary to design a climate-responsive building. A climate classification system based on the relationship between the climate variables and human thermal comfort would be more useful than a system based on vegetation or precipitation. Comfort depends on the dissipation of the body's metabolic heat production by radiation, convection, evaporation, and conduction; discomfort will occur when there is insufficient or too much dissipation. As discussed in Chapter 3, the amount of heat dissipated through various mechanisms varies, but, in still air and in temperatures close to 20°C, without conductive contact the heat dissipation will be about 45% by radiation, 30% by convection, and 25% by evaporation. This heat transfer exchange is affected by climate variables in different ways. The convective exchange is affected by the dry-bulb temperature and air velocity. The radiant exchange depends on the thermal relationship between the skin and the surfaces around it. The temperature of the skin is relatively constant, but the temperature of

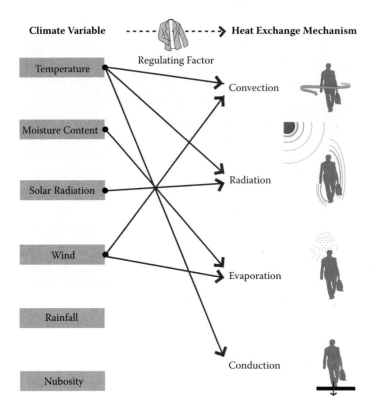

FIGURE 4.4 Effects of climate variables on heat transfer mechanisms.

the surrounding surfaces is affected by the temperature of the air and the intensity of solar radiation hitting these surfaces. The moisture content of the air affects the rate of evaporation from the lungs and the rate of sweat evaporation, which is also affected by air velocity. Conductive exchanges are usually minimal, because they occur only when the human body is in contact with a surface and are affected by the air temperature and radiation on the body or the surfaces (Figure 4.4).

Using this classification based on the connection between the climate variables and thermal comfort, climates can be sorted into four types: cold, temperate, hot and dry, and hot and humid. This simple climate classification is easier to understand and more practical for architectural purposes than Köppen's classification and has been adopted by many authors (Moffat and Schiler, 1981; Achard and Gicquel, 1986). It is possible to determine the type of climate by the location of the hourly or monthly data points in the psychrometric chart. This will be discussed later in the chapter.

The main problem in a cold climate is the lack of heat (i.e., underheating), which causes excessive heat dissipation during all or most parts of the year. In a temperate climate the problem is excessive heat dissipation during the warmest months of the year and inadequate heat dissipation during the coolest months of the year. There is a seasonal variation between underheating and overheating, but neither of the seasons is very severe. A hot, dry (arid) climate is typified by large diurnal temperature swings and the main problem is overheating, with insufficient heat dissipation from the body; however, the dryness of the air allows evaporative cooling to take place, which provides additional cooling opportunities for the body. Elevated temperature and radiation characterize these climates. In hot, humid climates the air temperatures are lower than in hot, dry areas, but there is relatively little night drop, and the main problem is still overheating with conditions aggravated by high humidity, which limits the evaporative cooling effect. If these four climate types are compared with Köppen's classification, a type A climate would be hot and humid, type B would be hot and dry, type G would be temperate, and type H would be cold.

4.4 CLIMATE ZONES AND ENERGY CODES

Legislation is an important tool for change. Unfortunately, advances in legislation are slow and should move to improve performance and away from prescriptive requirements to performance-based requirements based on indicators such as yearly energy use or greenhouse gas (GHG) emissions per unit area per unit of time. To achieve this, designers must be better educated in the available mechanisms. Even if countries sign accords to reduce emissions, it is impossible to reach these goals if codes do not mandate improved performance that will reduce emissions in all areas.

In addition to the major humid, dry, and marine climate types, the divisions among climate zones for energy codes have mostly been developed based on thermal criteria. After the 1980s, climate data were widely available in many locations in the United States, and computer-processing power was cheaper. This led to a climate classification based on the agglomeration of similar sites using a statistical procedure called *hierarchical cluster analysis*. This type of analysis uses a distance metric that represents the degree of similarity or dissimilarity between observations in a data set. Cluster analysis has been discussed by several authors (e.g., Oliver, 1991) and has been used by some standards such as ASHRAE Standard 90.1.

In the United States, probably the most important energy code is the International Energy Conservation Code (IECC), which covers new construction, additions remodeling, window replacement, and building repairs. The first IECC climate zones were based only on heating degree days and did not account for cooling energy (Building Energy Codes Resource Center, 2010). On the newer 2009 IECC maps, the climate zones were developed from the analysis of 4,775 National Oceanic and Atmospheric Administration (NOAA) weather sites. A climate classification system that provided quantitative definitions for three major climate types—humid, dry, and marine climates was developed. The reasoning is that these major climate types are the most essential for building design because they are based on the most important variables in cooling and heating loads, such as wet-bulb temperature and variations in solar radiation and heating and cooling degree days.

These three climate types are divided in eight climate zones that are defined by county boundaries and thus have been adopted by many organizations and codes including ASHRAE 90.1, ASHRAE 90.2, ASHRAE's Advanced Energy Design Guide for Small Office Buildings, Building America (modified), and Energy Star. The eight climate zones are distributed in roughly horizontal bands, except for the

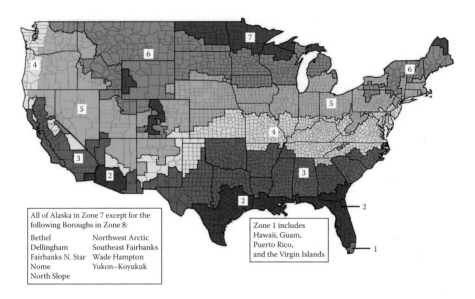

FIGURE 4.5 IECC climate zones in the United States.

Rocky Mountains, spread across the country from the hottest zone (1) occupying the southern tip of Florida to the coldest zone (8) covering the northern half of Alaska. Regular increments of heating and cooling degree days were selected that worked well climatologically, that facilitated the representation of current code requirements, and that enabled a substantial reduction in the number of zones (Figure 4.5).

California's Energy Efficiency Standards for Residential and Nonresidential Buildings were established in 1978 in response to a legislative mandate to reduce California's energy consumption and are probably the most stringent energy codes in the United States. According to the California Energy Commission (CEC), California's building efficiency standards (along with those for energy efficient appliances) have saved more than $56 billion in electricity and natural gas costs since 1978, and it is estimated that they will save an additional $23 billion by 2013. The standards are updated periodically to allow consideration and incorporation of new energy efficiency technologies and methods. The latest version of the standard went into effect in January 1, 2010. This code divides the state into 16 climate zones (Figure 4.6), each of which represents a geographic area with a similar energy budget, which is the maximum amount of energy that a building or portion of a building can be designed to consume per year (Title 24, Energy Efficiency Standards). The Energy Commission developed weather data for each climate zone by using unmodified weather year data for a representative city, which simplifies calculations but produces less precise numbers in some locations inside each climate zone. As with all climate zones, the borders sometimes are set by political rather than geographical boundaries, which makes enforcement easier but also generates problems with the precision of the budget.

The purpose of the European Union Energy Performance of Buildings Directive (EPBD) (2002/91/EC) is to promote the improvement of energy efficiency in

Climate and Architecture 105

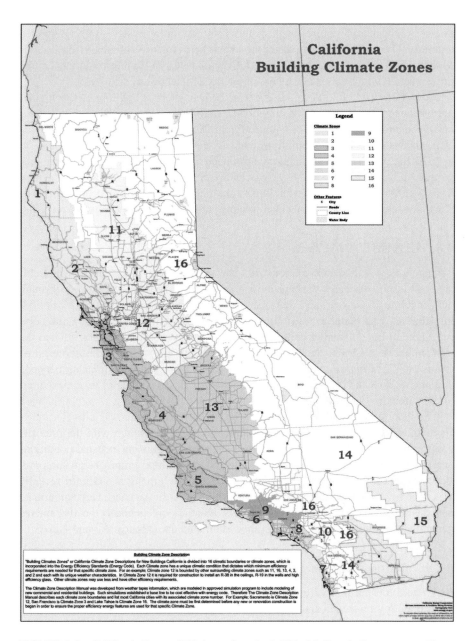

FIGURE 4.6 California Energy Efficiency Standards Title 24, Part 6 climate zones. (From California Energy Commission. With permission.)

buildings and to ensure convergence of standards across Europe. The law applies to both new projects and refurbishments and establishes a common basis for calculating energy performance, sets minimum standards, introduces energy performance certificates (EPCs), and requires heating and air-conditioning systems to be inspected regularly. The establishment of a common basis helps to unify all measuring systems so that the display energy certificates (DECs) or EPCs in smaller buildings can be implemented, which raises awareness of energy efficiency among users.

Many European building codes, such as the Danish 2008 building code, measure performance in total energy consumption in $kWh/m^2/yr$. The Danish code regulates new residential buildings at 70 $kWh/m^2/yr$ in 2008 and down to 25 $kWh/m^2/yr$ in 2020 and for nonresidential buildings from 95 $kWh/m^2/yr$ in 2008 to 35 $kWh/m^2/yr$ in 2020. In 2008 the United Kingdom introduced the Climate Change Act, which proposed to reduce CO_2 emissions by at least 80% from 1990 levels by 2050 and by at least 34% by 2020. This is a legally binding target that at the same time helps the United Kingdom achieve the reductions in emissions proposed in the 1997 Kyoto Protocol.

4.5 CLIMATE ANALYSIS

Climate analysis uses information from the building site to determine the architectural design strategies that are most effective to harness natural forces through building design and to achieve thermal comfort. This process helps to produce buildings that respond better to local climate, that use less energy for heating and cooling, and that have a smaller carbon footprint. The first modern text that studied the relation between climate and buildings was probably Koenigsberger, Mahoney, and Evans's (1971) *Climate and House Design*, which proposed a relationship between climatic regions and the recommended bioclimatic design strategies to achieve thermal comfort in each of these.

The next step in the development of bioclimatic strategies was climate analysis methods that compared average climatic data for each month with the required comfort levels to determine the difference between the existing external conditions and the desired indoor conditions and to decide on the appropriate design strategies to achieve thermal comfort. This analysis was usually graphical and until recently was done by manually plotting average monthly values in the chart. Current climate analysis software can now analyze hundreds of data points in hourly weather files to generate information that is easier to understand and use (Figures 4.2 and 4.3).

4.5.1 BUILDING BIOCLIMATIC CHART

Bioclimatic design is used to identify potential building design strategies that use natural energy resources and minimize conventional energy usage to achieve thermal comfort. The building bioclimatic chart indicates design strategies that can be used for given outdoor combinations of temperature and relative humidity to achieve indoor thermal comfort, and it is called *bioclimatic* because it connects climate with human thermal comfort. The first building bioclimatic chart to become generally accepted was by Olgyay, developed in 1963. In this chart, relative humidity is measured on the *x*-axis and temperature on the *y*-axis The comfort zone is plotted in the

Climate and Architecture

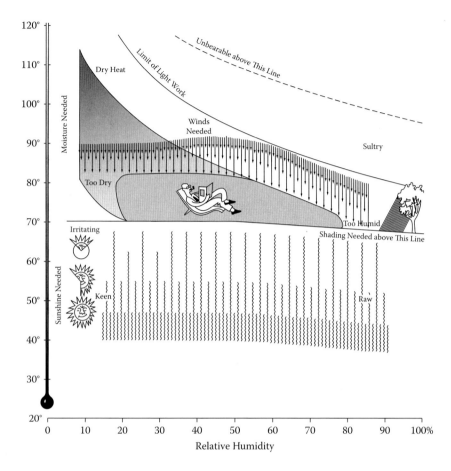

FIGURE 4.7 Olgyay's building bioclimatic chart. (From Olgyay, V., *Design with Climate, A Bioclimatic Approach to Architectural Regionalism.* Princeton, NJ: Princeton University Press, 1963. With permission.)

chart with a lower limit of the zone around 21°C (70°F) and an upper limit around 27°C (80°F), which gradually goes down until it connects with the lower limit of the comfort zone at a relative humidity of 90% (Figure 4.7). In addition to the comfort zone, this chart indicates wind and solar control strategies for different outdoor conditions expressed as combinations of dry-bulb temperature and relative humidity. Areas above the comfort zone are overheated and need ventilation, and areas below the comfort zone are overcooled and need solar radiation. Many versions of this chart have been developed and are still in use in different parts of the world; some of them indicate required air velocities or amounts of solar radiation to achieve comfort, which is probably the biggest strength of this chart.

An important problem with Olgyay's (1963) chart is that it does not account for differences between low-mass and high-mass buildings and assumes that the outdoor conditions, plotted on the graph, would be very close to the indoor conditions and can thus be used as guidelines for building design. Givoni (1998) states that this is only a reasonable approximation in lightweight naturally ventilated buildings in the summer of temperate or humid climates. But even in these buildings, especially at night, internal gains can generate winter indoor temperatures that can be significantly higher than outdoor temperatures, leading to an overestimation of the need for heating. In warm, dry regions, the daytime temperature inside a night-ventilated non-air-conditioned high-mass building can be several degrees lower than the outdoors. In this case, guidelines based on outdoor conditions would be inappropriate because the indoor temperature would be very different (lower) from the outdoor temperature, and guidelines based on outdoor values would recommend that the building be ventilated during the daytime and heated during the night, which would only overheat a high-mass building in the desert.

4.5.2 Givoni's Building Bioclimatic Chart

Givoni (1969) developed the building bioclimatic chart (BBCC) to address the problems associated with Olgyay's charts. This chart is based on the temperatures inside buildings (expected on the basis of experience or calculations) instead of the outdoor temperatures. Givoni used the psychrometric chart as the base of his BBCC for plotting the comfort zone and its extension (Figure 4.8) (Givoni, 1969). The psychrometric chart graphically represents the interrelation of air temperature and moisture content and the BBCC, which is drawn in the psychrometric chart, becomes a useful

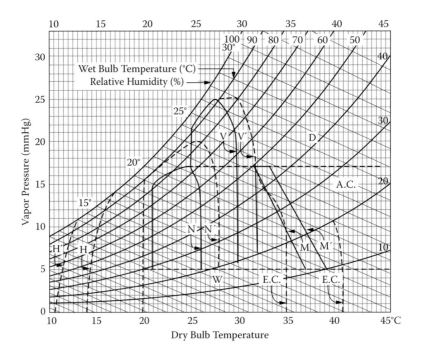

FIGURE 4.8 Givoni's building bioclimatic chart. (From Givoni, B., *Man Climate and Architecture,* London, Applied Science Publishers, 1969. With permission.)

Climate and Architecture

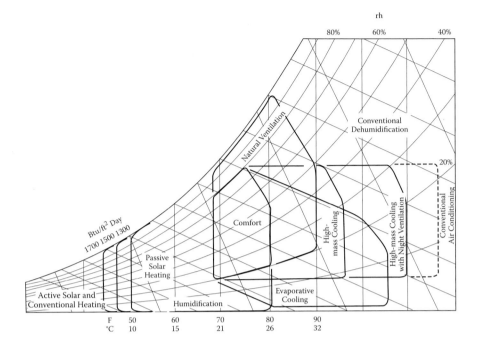

FIGURE 4.9 Milne and Givoni's revised building bioclimatic chart. (From Milne, M. and Givoni, B., in *Energy Conservation through Building Design*, D. Watson (Ed.), New York, McGraw Hill, 96–113, 1979. With permission.)

design tool that indicates the boundaries of the outdoor climatic conditions within which various passive building design strategies can provide indoor comfort without the use of mechanical systems. In Givoni's original chart, several zones were listed together with the comfort zone. These indicate when different strategies can be used to achieve thermal comfort: natural ventilation, thermal mass, evaporative cooling, and air conditioning. The strategies are the same, but research has changed the demarcation of the zones. Milne and Givoni's (1979) chapter in *Energy Conservation through Building Design* added heating zones and further developed the method (Figure 4.9). Again, the cooling options included in the chart are daytime ventilation, high mass with or without nocturnal ventilation, direct evaporative cooling, and indirect evaporative cooling. Heating is addressed through passive solar heating. This chart and versions of this chart with different demarcations of the zones are used all over the world. Gonzalez, Hinz, Oteiza, and Quiros (1986) compared several of these building bioclimatic charts to determine a new chart for hot and humid climates. These strategies are discussed in Chapter 6.

In 1998 Givoni revised the original chart based on updated research and proposed a revised BBCC, which also included a distinction between developed and developing countries. In these charts he assumed that the upper limits of accepted temperature and humidity would be higher for persons living in developing countries and acclimatized to hot, humid conditions and extended the zone by which air movement could provide comfort by a couple of degrees to the right (Figure 4.10). To support

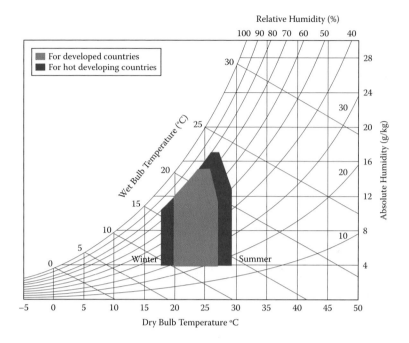

FIGURE 4.10 Givoni's comfort zones for groups in developed and hot developing countries.

this reasoning he used data obtained in comfort studies conducted in hot countries (Givoni, 1998). However, as stated earlier, social, cultural, and economic factors such as income level and work and home environments still play an important and unaccounted for role that could generate different comfort zones among different social groups, even in the same location. For example, an individual living in a hot and humid climate and working and traveling in an air-conditioned environment would have a different comfort zone from an individual in the same city living, traveling, and working in non-air-conditioned environments. In this case, the zones for less developed and developed countries could be indicating zones for different groups inside countries instead of whole countries.

In the same fashion, Givoni also proposed a larger and warmer zone in which ventilation will help to achieve thermal comfort in hot developing countries (Figure 4.11).

The BBC plotted in Figure 4.12 is the author's version of Milne and Givoni's (1979) chart and will be used to explain passive solar concepts in Chapter 6. It follows Milne and Givoni's zones with two exceptions. First, the internal gains strategy is eliminated, because it is not really an architectural design strategy. It is more affected by the building program and in many cases is a liability instead of a strategy. Second, the thermal mass area that is traditionally to the right is integrated into one zone with thermal mass to the left and the right of the comfort zone. This indicates that the thermal mass can dampen outdoor swing when outdoor values are slightly toward either side of the comfort zone (cooler or warmer), bringing them closer to thermal comfort. As building technology evolves and improves, the boundaries of all the strategy zones

Climate and Architecture

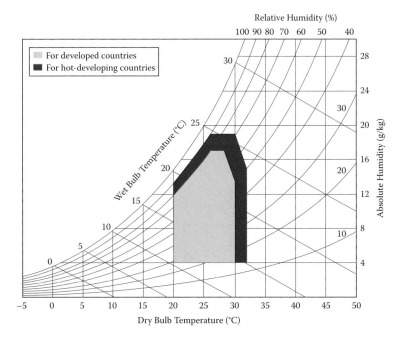

FIGURE 4.11 Givoni's BBC for comfort ventilation in developed and hot developing countries.

will extend farther and farther away from the comfort zone, and it will be possible to design climate-responsive low-energy buildings under more difficult conditions.

Conditions toward the left of the comfort zone are cooler than comfort, indicating that heating must be implemented to achieve comfort. Conditions toward the right of the comfort zone will require implementation of a cooling strategy. As the points are farther away from the comfort zone in either direction, more heating and cooling are required. After a certain point it is not possible to heat or cool the building solely with passive means and mechanical systems will be implemented.

4.5.3 Digital Climate Analysis Tools

As described in Section 4.2, digital climate analysis tools can process yearly data that are available in several weather file formats such as EPW and weather (WEA) to present data graphically so that it is easier to process and understand, even though it is statistical, not real. The ability to process yearly data files is the main advantage of these digital tools; however, it is also their main limitation. If a yearly climate data file is not available, then it is not possible to do any analysis at all unless data are interpolated from nearby stations or data are used from a similar location and organized in a readable form. Weather analytics and Meteonorm can generate weather files for different locations by interpolating existing weather files. Rodriguez (2008), a professor from Universidad del Zulia in Venezuela, developed a simple climate analysis tool called Clima that cannot process annual climate files, but it has the

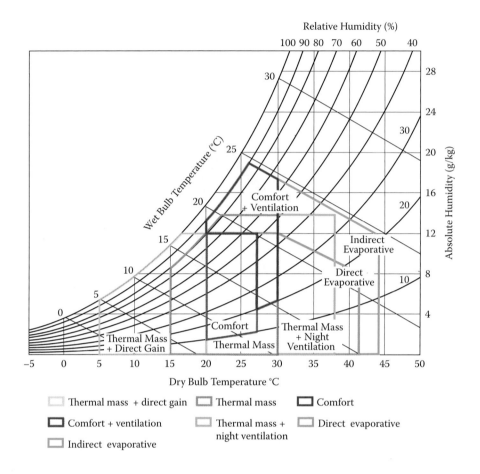

FIGURE 4.12 Proposed building bioclimatic chart with strategies.

advantage of processing manually introduced data (minimum and maximum average monthly values) providing more opportunities to process information from any location in the world, which it plots in a psychrometric chart using Givoni's comfort zones for developed and less developed countries.

All of these newer digital tools incorporate the use of concepts such as the BBCC and the timetable of bioclimatic needs. The latest version of Climate Consultant (Milne, 2011) also provides multiple comfort models such as the California Energy Code comfort model for 2008, the *ASHRAE Handbook of Fundamentals* comfort model (2005), ASHRAE Standard 55-2004 using the predicted mean vote (PMV), and the adaptive comfort standard in ASHRAE Standard 55-2004. In addition, it provides many useful options to visualize and better understand data. Figure 4.13 shows a digital version of Olgyay's timetable of bioclimatic needs, plotted from Climate Consultant. Hourly temperatures are plotted in the table and are color coded according to their temperature range, which gives an idea visually of the yearly temperature range and the temperature location in one single image. Figure 4.14 shows a wind

Climate and Architecture

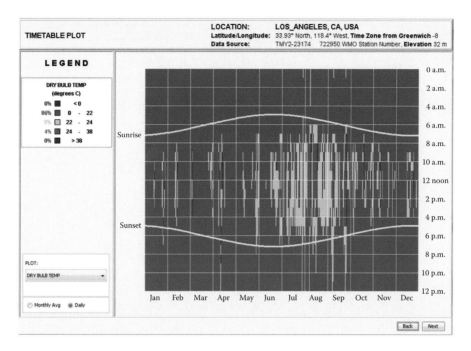

FIGURE 4.13 Timetable of bioclimatic needs from Climate Consultant.

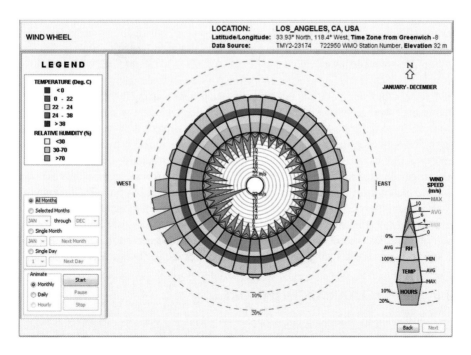

FIGURE 4.14 Wind wheel from Climate Consultant.

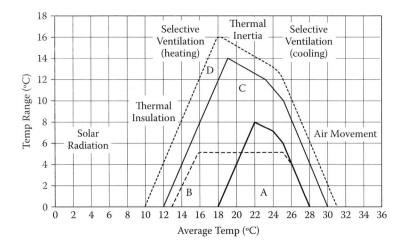

FIGURE 4.15 Comfort triangle charts. (From Evans, M. (2007). The Comfort Triangles: A New Tool for Bioclimatic Design. PhD thesis. Delf Technical University. With permission.)

wheel developed by Milne and Al-Shaali for Climate Consultant, which in addition to wind direction and frequency includes air temperature and relative humidity information and can be studied in yearly or monthly increments.

4.5.4 THE COMFORT TRIANGLES CHART

Evans (1998) developed another climate analysis tool called the comfort triangles chart. This climate analysis tool was originally proposed by Evans and de Schiller (1998) and then further developed by Evans (2003, 2007). In his chart, Evans also uses two variables to characterize conditions in a specific moment; however, instead of using dry-bulb temperature and humidity, he proposes using the daily temperature swing and average temperature because these characterize the temperature variations during a daily cycle. The average temperature is plotted on the x-axis, and the swing is plotted on the y-axis. The chart indicates the four different zones for sedentary activities, sleeping, circulation, and extended circulation in addition to the bioclimatic strategies that can be applied to improve comfort when the conditions are outside the comfort zones: sensible air movement; thermal inertia, which combines time lag and thermal dampening; solar radiation; and thermal insulation and selective ventilation (Figure 4.15). As with other climate analysis tools, the analysis of the outdoor data leads to design strategies that can be implemented to achieve thermal comfort. Evans (2007) validated his system in several climates with positive results.

4.6 VERNACULAR ARCHITECTURE

Even though climatic determinism fails to account for the range and diversity of house forms, climate is, nevertheless, an important aspect of the form-generating forces, and has major effects on the forms man may wish to create for himself. This is to be expected under conditions of weak technology and limited environmental control

Climate and Architecture

systems, where man cannot dominate nature but must adapt to it…. The principal aspect to be examined is the amazing skill shown by primitive and peasant builders in dealing with climatic problems, and their ability to use minimum resources for maximum comfort." (Rapoport, 1969, p. 40)

Primitive and preindustrial builders learned to solve their problems by collaborating with nature. The buildings were designed to protect their way of life. If the building failed, then the owner would be exposed to the potentially harsh forces of nature. These buildings show respect for the environment, and it is possible to learn from them. It could even be argued that the bioclimatic concept, and the now more global and comprehensive notion of sustainable architecture, extend with scientific methods the ecologically responsible concepts transmitted through many generations, based on an intuitive knowledge of the environment and the climate.

4.6.1 Vernacular Architecture in Warm, Humid Climates

In a warm and humid climate, overheating is not as significant as in hot and dry areas, but conditions are aggravated by high humidity, which limits the evaporative cooling effect. These climates have at least 1 month of the year with a temperature higher than 20°C and a mean relative humidity around 80%. The monthly rainfall is often above 200 mm, mostly in the form of short-duration, high-intensity showers.

A vernacular building in a hot and humid climate is usually lightweight, very open to the winds to promote evaporative cooling, and with a large roof that extends over the walls to block solar radiation (Figure 4.16). There are many variations of this model, but vernacular buildings like this can be found in hot, humid climates all over the world, from the South Pacific to the Caribbean. Because it was so important

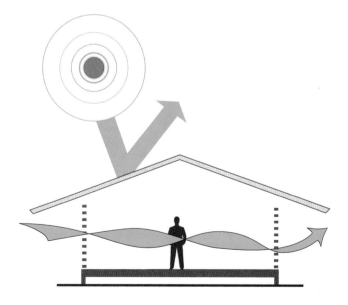

FIGURE 4.16 Hot and humid climate archetype.

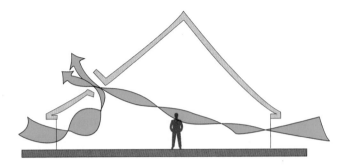

FIGURE 4.17 Yekuana churuata.

to provide natural ventilation to these buildings, they were usually spread apart so that they would not block breezes from each other.

The Churuata is the typical dwelling of several ethnic groups in southern Venezuela, both in the Gran Sabana and the Amazon region. The churuatas of the Piaroa, Ye'kuana, Hoti, Panare, and Pemon are similar but not identical; each ethnic group has its own characteristic construction, all with great formal beauty (Gasparini, 1993). The Piaroa have a tendency to elevate the height of the conical roof, and some have oblong plans with rounded ends. Usually the bottom of the churuata consists of a peripherical mud wall and a thatched roof that accounts for most of the building envelope (Figure 4.17). The bottom wall has enough shaded thermal mass to modulate temperature, while the upper roof provides shade and stores humidity from the air, lowering the air temperature by evaporation. Most of the churuatas have an opening in the roof for sunlight and evacuation of hot air (Figure 4.17). Interior compartments give some privacy to its dwellers.

Many Native American communities developed their settlements by the riverbanks or lakes, which were used for communication. These dwellings were raised above the water, on pilings, and construction elements were simplified to the extent that many of their houses simply became a roof on top of wooden poles. Other environmental factors such as the height of the maximum tides and floods were also taken into consideration in their design. An example is the dwellings of the Paraujanos in the Maracaibo Lake Basin in western Venezuela. These buildings perform as an inhabited filter in which the different components of the envelope—floors, walls, and ceilings—are permeable; sunlight, wind, dust, and water are regulated by these three components (Machado, La Roche, Mustieles, and De Oteiza, 2000). The floor, built with mangrove cane poles, permits cool air from the water below to filter inside the building. The walls, also made of cane poles or thatch, are permeable to the light and wind, filtering in the light (Figures 4.18 and 4.19). The ceiling, thick and built with enea, a type of thatch, allows the house to breathe and evacuates the heat but blocks the rain. Because it is opaque, enea also blocks most of the solar radiation to the interior of the building, effectively shading the interior. This type of house is still built today, even in urban settings, but unfortunately many of the original materials have been substituted with contemporary materials such as industrialized tin and zinc, with a longer life span but much reduced performance.

Climate and Architecture

FIGURE 4.18 Paraujano house.

FIGURE 4.19 Paraujano house. (Photo courtesy of Carlos Quiros.)

Some contemporary proposals in the Sinamaica Lagoon, such as the community center and church designed by the author and built by the Corporation for the Development of the State of Zulia, considered local architecture and materials. In the meeting room, for example, the air filters through the mangrove poles, providing natural ventilation with air that is cooled as it passes above the water. These poles are further separated above the door lintel to increase airflow inside but to maintain a certain privacy at the lower level.

An interesting example of nonvernacular architecture adapted to a warm and humid climate is the architecture for oil and fruit companies of the early to mid-20th century. This architecture was the result of the research and experience of

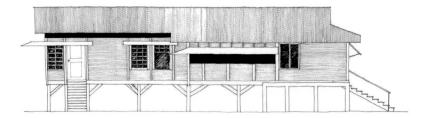

FIGURE 4.20 Standard Oil Company house in Maracaibo, Venezuela. (Drawing courtesy of Maria Gabriela Romero.)

FIGURE 4.21 Standard Oil Company house in Maracaibo, Venezuela. (Photo courtesy of Pedro Romero.)

Dutch, British, and American transnational corporations sometimes associated with researchers and universities, and many variations of these buildings can still be seen all over the world. These houses were usually detached rectangular buildings, completely surrounded by gardens, and most often with light-colored facades, with their windows and doors protected by a large roof or "big hat." The roof generally had four slopes, with metal sheets and a ventilated attic, above a wooden or metal structure (Figures 4.20 and 4.21). The exterior walls were usually made of wood or metal, sometimes with an air space, and provided relatively good insulation for their time (La Roche, 1992). These homes, which were designed to be climate responsive, have continued to evolve and generate a local vernacular language in many regions of the world. Figure 4.22 shows one of these in Panama, while Figure 4.23 shows another one in Kuai, Hawaii.

Climate and Architecture

FIGURE 4.22 Hot, humid contemporary vernacular in Panama.

FIGURE 4.23 Hot, humid contemporary vernacular in Kuai.

4.6.2 Vernacular Architecture in Warm and Dry Climates

The main problem in a hot, dry climate is overheating (i.e., inadequate heat dissipation from the body), but because the air is dry, evaporative cooling can take place. High temperatures and radiation characterize such climates, and the mean temperature of the warmest months is above 27°C with a low relative humidity. The yearly maximum temperature is around 45°C, and the minimum can be as low as –10°C, which is a large temperature swing.

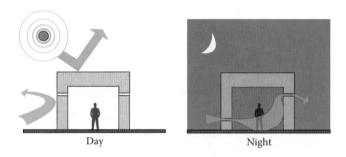

FIGURE 4.24 Hot and dry climate archetype.

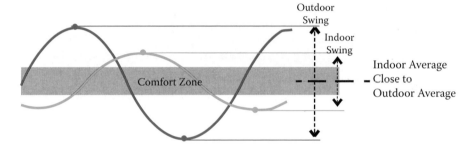

FIGURE 4.25 Effect of thermal mass on indoor daily temperature swing.

A building in a warm and dry climate is usually a compact heavy mass structure that modulates outdoor temperatures, with small windows to reduce solar gains to the interior of the space but that allow some ventilation for night cooling (Figure 4.24). Because of the need to reduce solar gains, the building layout is also compact, reducing the amount of exterior surface exposed to the sun, receiving direct solar radiation. Controlling natural ventilation allows the occupant to capture night-time temperatures in the interior thermal mass, cooling the interior of the space, which then acts as a heat sink during the daytime absorbing heat and keeping it cool through the day. The effect of thermal mass by itself reduces the daily indoor swing (Figure 4.25) so that indoor temperatures are not as extreme, and the resulting indoor temperature fluctuates somewhere close to the average of the outdoor temperature. When the user has the ability to control ventilation and night flushing is provided, the indoor temperature swing is still lower than the exterior swing but the resulting indoor temperature is shifted lower and closer to the outdoor minimum temperature (Figure 4.26). The reduction of the indoor temperature is a function of the amount of thermal mass, the air change rate during the night, and solar control during the day, usually as insulation, shading and reduced infiltration.

The *hatos*, or small farms, of the Paraguana Peninsula in Venezuela (Figure 4.25) are generally compact to reduce surface area exposed to solar radiation and to increase the amount of mass per dwelling and thus thermal inertia, which flattens temperature swing. The exterior color of these hatos is usually white, which reflects

Climate and Architecture

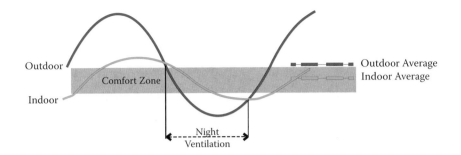

FIGURE 4.26 Effect of thermal mass and night ventilation on indoor daily temperature swing.

FIGURE 4.27 Hato in Falcon.

a large amount of incident solar radiation. Openings toward the outside are few and small, and there might be an outdoor transition space. The control of openings is very important: They must be closed during the day to avoid heat gains by radiation and convection and open during the night to permit the entry of cooler air that will lower the temperature of the mass inside the hato. Most of these houses were built using either adobe bricks or bahareque walls as the primary material for the walls, both of which have high thermal inertia. This mass also generates thermal lag with indoor temperatures that are sometimes uncomfortable during the night. Hatos are sometimes open to planted courtyards, and the kitchen is usually toward the southwest, permitting the northeastern tradewinds to carry heat, smoke, and smells away from the house.

In Andalucia, Spain, compact buildings are organized around courtyards that are not only beautiful, but also help to regulate climate by controlling radiant, convective, and conductive gains and losses (Reynolds, 1992). Shading to block solar heat gains is usually accomplished with toldos, moveable fabrics with varying degrees of translucency that block most of the direct solar radiation but permit diffuse light to penetrate (Figure 4.28), and some direct light that generates lively shadow patterns. These toldos can be opened at night to promote radiant cooling to the night sky from the courtyard, especially from the sky-facing floor. Some patios also use trellises and vines for shading but these do not have the operability of the toldo. The mass in the courtyards is a thermal regulator that reduces daily temperature

FIGURE 4.28 Courtyard in Cordoba with Fountain and Toldo. (Photo by John Reynolds, reprinted by permission from Reynolds, John S., *Courtyards: Aesthetic, Social, and Thermal Delight*, John Wiley & Sons, Hoboken, NJ, 2002.

swings. This mass, especially the sky-facing floor, is cooled at night by radiation to the night sky, and also by evaporation and ventilation. During the daytime this thermal mass then acts as a heat sink delaying, with the aid of the toldos, the arrival of the afternoon heat. By their radiant effect these cool surfaces also contribute to the comfort sensation. Plants also have an important role in these courtyards, they are not only a delight to the senses, they also demand caretaking and watering, which then provide through evaporation a major source of courtyard cooling in hot weather (Figure 4.29). Fountains and other bodies of water in the courtyard also provide additional opportunities for evaporative cooling.

Orientation and additional geometrical properties also affect thermal performance of these courtyards. The aspect ratio, which is the area of the courtyard floor to the average height of the surrounding walls, determines the degree of openness to the sky. The greater the aspect ratio, the more exposed this courtyard is to the sky allowing more heating during the day, and more floor surface facing the sky that can

FIGURE 4.29 Potted plants on patio in Cordoba. (Photo courtesy of Lucia La Roche.)

be cooled at night, and also more possibility for wind entry. The solar shadow index is the south wall height over the north–south floor and determines the amount of winter sun exposure. A higher index means that the courtyard well is deeper and less winter sun will reach its floor. There are many different combinations of orientation and aspect ratios and the solar shadow index that affect how the courtyard receives solar radiation.

Different sources define temperate climates in different ways. The definition stating that a temperate climate has some heating and cooling requirements but that neither is very severe is the one that is used. However, there are many variations of this and some temperate climates may have both a winter and a summer with rather severe variations in each. This is especially true in mid latitudes and inland of large continents such as North America and Asia (Moffat and Schiler, 1981). The mildest form of a temperate climate is modulated by masses of water and is the climate of the Mediterranean, much of the California coast, and many Latin American cities that are further away from the equator or at a higher altitude.

In inland temperate climates, a building has to respond to relatively warm summers and cold winters. Thus, the building must keep the heat inside in the winter, protecting from the cold winds while providing daylight. During the summer, the building must keep the heat outside, while also providing shade and ventilation. This means that an archetype in this climate should be compact, well insulated, and tight. In the coastal or milder areas the outdoor conditions are much milder with less temperature differences between the summer and the winter. Therefore, there is more flexibility in the design of the building as long as it provides shade and opens to the cool summer breezes in the summer, and protects from the cold wind, while providing some passive solar heating in the winter (Figure 4.30).

4.6.3 Vernacular Architecture in Temperate Climates

A temperate climate usually has some heating and cooling requirements, but neither is very severe. This is the climate of the Mediterranean, much of the California coast, and many Latin American cities that are farther away from the equator or at a higher altitude. There is no real archetype for a temperate climate because it is so forgiving that almost anything can be done as long as the building opens to the cool summer breezes and closes to the colder winter air while providing some passive solar heating (Figure 4.30). The overhangs and layout of the building should be adapted to latitude and provide solar radiation and shade as needed.

The Spaniards arrived in America in the late 15th century with a housing typology that incorporated principles of traditional southern Spanish dwellings. This model was implemented with variations adapted to different climates. Variations included building massing and orientation, window size and position, layout of the building, verandas, courtyards, and interior layout of the spaces. Most Latin American colonial architecture in temperate climates includes extensive use of verandas and courtyards (Figure 4.31). Many of these buildings such as the *haciendas*, or large farms, had large windows to promote ventilation and shaded external corridors in contact with gardens. Convents and monasteries were organized around courtyards for daylight, natural ventilation, and even collection of water (Figure 4.32).

Colonial residential buildings also had one or more courtyards for natural ventilation and illumination. At the front of the house toward the street, there usually was a bedroom and a living room, while the entrance was through a hallway called the *zaguan*. The dining room was toward the middle; the kitchen, laundry, and a multipurpose room sat toward the back and opened to an L-shaped courtyard.

There are many qualities worth preserving in Spanish colonial architecture; however, one that is often overlooked is the colonial window, which is a very good example of how windows can combine different components to provide different functions (Figure 4.33). Venetian blinds or wood louvers reduce sunlight as well as the intensity of light. They do not retain heat or let it go inside, yet they let air and

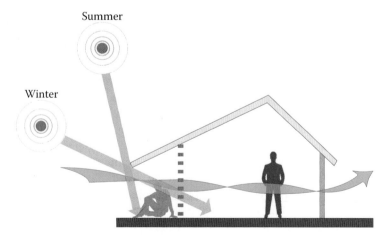

FIGURE 4.30 Temperate climate archetype.

FIGURE 4.31 Veranda in Quinta Anauco, Venezuela. (Courtesy of Carlos Quiros.)

breeze slip through. They also preserve privacy: You can look out without being seen (Figure 4.34). Waterproof wooden shutters protect from unwanted rain, wind, or dust; in many cases they have a small upper glass window that provides the inside with proper lighting even in the most adverse weather. Wooden bars separated from each other by no more than 15 centimeters provide security. Because the windows protrude outward, communication with bystanders or neighbors is promoted thanks to an indoor bench (Machado et al., 2000).

Another interesting example of climate-responsive buildings is the Case Study Houses in Los Angeles. The Case Study House program was sponsored by John Entenza's *Art & Architecture* magazine between 1945 and 1966. During this period, 36 experimental prototypes were designed and many of them built. These are not vernacular because they were designed by architects; however, even if the words *sustainability* or *green architecture* did not exist, these buildings already exhibited many of the attributes of contemporary green buildings. Ester McCoy (1989) stated the importance of climate in the design of these houses: "It was the California sun rather than the hearth that was at the base of the Case Study Houses." In most of the Case Study Houses the barrier between man and nature was broken; the houses were one with nature. The large operable windows and doors not only were open to the

FIGURE 4.32 Courtyard in Mexico used for daylight and natural ventilation.

views but also permitted the flow of air in the summer while permitting solar radiation in winter (Figures 4.35 and 4.36). The climate-responsive strategy of allowing free flow of space inside the modernist architecture of the Case Study Houses was an architectural design strategy to make small buildings feel larger (La Roche, 2003; La Roche and Almodóvar, 2009). The ability to ventilate, combined with summer shading and good site design, made most of these Case Study Houses effective in the summer, but the large amount of single-pane windows did not permit them to perform well in the winter (Figure 4.36).

The strong relationship of most of the Case Study Houses with the environment established an implicit adaptation with the local climate. But there were technological advances compared with vernacular buildings; most of the Case Study architects exploited the available materials to their full technological potential. Sustainability is not about imitation and nostalgia; it is about advancing design as it continues to evolve. So even though the Case Study Houses were innovative at one point, they are already obsolete in their use of materials. Current green buildings have many technological advances not available at that time. However, many of the principles of the Case Study Houses are still valid and will probably continue to be for generations.

Climate and Architecture

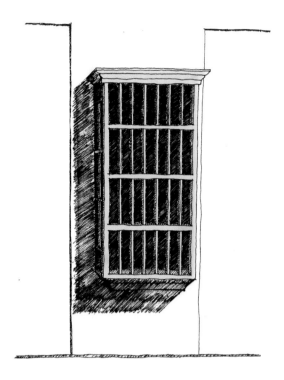

FIGURE 4.33 Typical Spanish colonial window in Maracaibo, Venezuela.

FIGURE 4.34 Colonial facade with several windows in Maracaibo, Venezuela. (Courtesy of R. Cuberos.)

FIGURE 4.35 View toward the south in Case Study House 22.

FIGURE 4.36 Case Study House 22.

4.6.4 VERNACULAR ARCHITECTURE IN THE COLD CLIMATES

In cold climates, the main problem is the lack of heat (i.e., underheating), which causes excessive heat dissipation for all or most parts of the year. A climate-responsive building in a cold climate usually has an envelope that covers a large volume with a small amount of skin. This is one of the reasons that many vernacular buildings in cold climates are variations of domes, cylinder or cubes, built with different materials. Insulation is usually provided by the thickness of the walls, which are generally built using local materials such as rock, earth, wood, or even snow. Windows are small to avoid heat losses, even though this also means reducing the amount of radiation and light entering the building. These buildings usually also have an internal heat source, such as a fireplace (Figure 4.37).

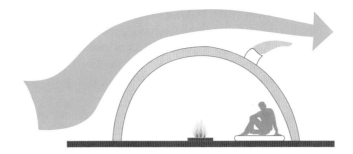

FIGURE 4.37 Cold climate archetype.

FIGURE 4.38 Vernacular building in the Andes Mountains in Venezuela.

There are many vernacular variations of buildings in cold climates. Figure 4.38 shows a vernacular building in the Venezuelan Andes Mountains. It is also compact with very few windows to the exterior. The exterior walls shelter from the cool outdoor winds, while a fireplace warms the interior of the space.

4.7 EFFECTS OF CLIMATE ON EMISSIONS

Vernacular buildings were not designed to operate with mechanical systems, and they used architectural features only to achieve comfort. However, many of these original vernacular typologies have also been modified over time due to changes in standard of living or cultural factors.

Energy used for operation of the building is the single most important source of emissions, and the intensity of these emissions is affected by the climate of the site and the building's program. To determine the effect of climate on energy emissions, the energy use of a single-family dwelling was calculated in these four climates. To better determine the effect of climate on emissions, emissions from construction, waste, water, and transportation were kept constant.

The house used in the simulations was two stories high and 14.6 m by 7.3 m (48 by 24 ft.), with 214 sq mt (2,300 sq. ft.) gross building floor area, similar to the U.S. average house in 2005. Climate files for the four previously discussed climate types were selected: (1) hot and dry, represented by California climate zone 15 (El Centro); (2) hot and humid, represented by Miami, Florida; (3) temperate, represented by California climate zone 6 (Los Angeles); and (4) cold, represented by California climate zone 16 (Bishop, CA). Emissions from operation changed as the heating, ventilating, and air-conditioning (HVAC) system used more or less energy to provide thermal comfort in the different climates.

4.7.1 Assumptions for Simulations

California's energy code Title 24 (California Energy Efficiency Standards, 2008) was used to design the building envelope for all climate zones; package D for homes with natural gas was used. A gas furnace with an annual fuel utilization efficiency (AFUE) of 80% was used. The same envelope was used in the hot and humid zone as in the hot and dry climate zone (Table 4.1).

4.7.2 Operation

Operational emissions involve all the emissions produced from the energy used to keep the buildings and everything inside it running. According to Bordass (2006), the process to estimate CO_2 emissions generated from operation in buildings involves five steps:

1. Define the boundary of the premises. Boundaries should be where they make practical sense in terms of where the energy can be counted (e.g., the area fed by the meters) and how the area is run (a tenancy, a building, a site, or even a district or a city). One may look at more than one boundary: for a university, the campus, specific buildings, and individual departments; for a rented building, the whole building and each tenancy.
2. Measure the flows of each energy supply across the defined boundary. Normally this will be annual totals by fuel, though details of load profiles could sometimes be included.
3. Define carbon dioxide factors for each energy supply.
4. Multiply each energy flow by the appropriate carbon dioxide conversion factor to get the emissions associated with each fuel.
5. Add them up to obtain the annual total of CO_2 emissions.

TABLE 4.1
Envelope Characteristics

	Referenced City	Title 24 Package C				Title 24 Package D			
		Ceiling	Wood Frame Walls	Glazing U-Value	EGR (Glazing Max Area)	Ceiling	Wood Frame Walls	Glazing U-Value	EGR (Glazing Max Area)
California CZ 16 cold	Bishop	R49	R29	0.42	14%	R38	R21	0.55	20%
California CZ 15 hot and dry	El Centro	R49	R29	0.38	16%	R38	R21	0.55	20%
Miami hot and humid	Fresno	R49	R29	0.38	16%	R38	R19	0.57	20%
California CZ 6 temperate	Los Angeles	R38	R21	0.42	14%	R30	R13	0.57	20%

To calculate the CO_2 factors for each energy supply, the emission factors must be determined for each one. In the United States there are several methods to determine the conversion factor (Bryan and Trusty, 2008):

1. The U.S. Environmental Protection Agency (EPA) Power Profiler calculates CO_2 emission factors for historical yearly average emissions for every U.S. zip code.
2. The EPA eGRID is a database that has hourly CO_2 emissions for every U.S. power plant. However, these data are also historical, and it is a nontrivial task to estimate marginal generation from this database.
3. The National Renewable Energy Laboratory (NREL) model develops direct and indirect impacts for typical building fuels and uses CO_2e, which includes other important GHGs besides CO_2, like methane (CH_4) and nitrous oxide (N_2O). This model generates emission factors for all U.S. regions as well as the nation.
4. The CEC E3 model uses the output of a production simulation dispatch model to forecast average and marginal CO_2 emission factors for California.

Two programs, Home Energy Efficient Design (HEED) and Design Builder, were used to predict energy use in the four locations. HEED (Milne, 2007) is an energy analysis tool that calculates the building's performance. When the program is first launched, it asks four questions about the project (i.e., building type, square footage, number of stories, and climate location), and with this information it creates Scheme 1, a building that meets the California Energy Code. It then designs a second scheme that is usually about 30% better. Next, it suggests other strategies that designers can test using the remaining seven schemes. HEED makes it very easy for users to change any aspect of the building's design, and after each design change it shows how the building's performance compares with the initial schemes.

DesignBuilder is a user interface to the EnergyPlus dynamic thermal simulation engine. It features an Open Graphics Library (OpenGL) solid modeler, which allows building models to be assembled by positioning, stretching, and cutting blocks in three-dimensional (3-D) space. 3-D elements provide visual feedback of actual element thickness and room areas and volumes. Data templates allow users to load common building constructions, activities, HVAC, and lighting systems into the design by selecting from drop-down lists. These templates can also be added. Users can switch between "Model Edit View" and "Environmental Performance Data," which is displayed without needing to run external modules and import data.

The yearly energy used by the building as determined with these programs is multiplied by a conversion factor to determine GHG emissions. Because different locations use different utilities at different times, which would further muddle the numbers, the same conversion factors were used for electricity and gas in all the climates: 0.62 k of CO_2 per kWh for electricity, which is the average value for the United States (egrid, 2007); and 0.19 per kWh (5.43 k of CO_2 per therm) for natural gas, which is the value proposed by DEFRA (2005).

4.7.3 CONSTRUCTION

Even though most of a building's emissions originate from its operation, carbon neutrality is more than just the operation of the building and includes the building's life cycle, from fabrication to demolition. GHG emissions from construction processes are usually related to the embodied energy in the building and can be generated in three different distinct stages: (1) during the fabrication of the materials used in the building; (2) during transportation of materials to the building; and (3) during construction of the building. Materials that need more energy to produce, such as steel, aluminum, and cement, are responsible for more emissions than materials like stone or wood. Local materials also usually produce fewer emissions than materials that have to be transported from longer distances. The construction process affects the embodied emissions, and an efficient, streamlined semiprefabricated system will produce fewer emissions than an inefficient construction process.

Emissions for construction were calculated using buildcarbonneutral, a very simple calculator that provides rough results. The total emissions from construction were 53 metric tons of CO_2/year (116,812 lbs/yr). A building life of 50 years was assumed, which is equivalent to 1061 kg/yr of CO_2e (2336.2 lbs/yr) over the life of the building, and this same number was used in all sites. If construction components could be recycled or the life of the building was extended, then the impact would be lower. If the building required major renovations, then construction-related emissions would be higher. More precise data can be generated using Athena EcoCalculator for assemblies. However, this calculator does not provide results for all regions. Results from buildcarbonneutral are not precise, and even though the imprecision could be quite substantial, the results can provide a preliminary idea of construction-related emissions.

4.7.4 WASTE

Waste generated from the building must also be treated. More waste requires more treatment and can generate methane, which is a potent greenhouse gas produced in landfills. To figure out these emissions, the waste section of the carbon emissions calculator developed by the EPA was used (EPA, 2009). Even though many cities and households recycle, it is still not mandatory in many regions of the United States, so for these calculations I did not assume that the household recycled. According to this calculator, emissions from waste for a family of four would be 1856 kg CO_2e, and if recycling was implemented (e.g., plastic, aluminum, newspapers, glass, magazines), emissions would be reduced to 1044 kg CO_2e. If more recycling and composting were introduced, emissions would be reduced even more.

4.7.5 WATER

The process used to treat the water provided to the building and coming out of the building usually is also responsible for the generation of GHG emissions. The water used in the building must be pumped from the source, then treated to be made potable, and then pumped to the building for consumption. The wastewater from the building must also be treated, which also generates carbon emissions, indirectly by using energy for these processes and directly by releasing methane.

A study on water-related energy use in California (Commission on Energy, Water and Parks, 2007) calculated the embedded energy in water for Southern and Northern California. The study estimated the amount of energy needed for each sector of the water-use cycle in terms of the number of kilowatt-hours (kWh) needed to collect, extract, convey, treat, and distribute 1 million gallons (MG) of water and the number of kWh needed to treat and dispose of the same quantity of wastewater. For Southern California, the embedded energy per MG was estimated to be 13,021 kWh. To provide 552,670 liters (146,000 gallons) of water per year to a family of four in Southern California requires 1,901 kWh. If every kWh of electricity in California generates about 0.32 kg (0.7 lbs) of CO_2, then a family of four generates about 605 kg (1,331 lbs) of CO_2/year, which is equivalent to 0.0034 kg of CO_2/L. Again, to simplify, this number was used for all sites even though the number for Southern California is probably higher than in the rest of the United States.

4.7.6 Transportation

By its location, a building is also indirectly responsible for CO_2 emissions from transportation because it affects how building users move to and from the building. A location close to public transit lines or in urban areas with higher density and walkable neighborhoods usually reduces transportation-related carbon emissions. For this analysis, emissions from transportation were not included.

4.7.7 Carbon Emissions in the Four Climates

In the cold climate, the average emissions from operational energy are 6,050 kg of CO_2e, which is 69% of the total emissions (Table 4.2). To determine the impact of the different sources of operational energy, a more detailed analysis including heating, cooling, fans and blowers, appliances, water heaters, and lighting was performed. This analysis indicates that in the cold climate, after averaging the results

TABLE 4.2
Source of CO_2 Emissions in Different Climates

Emissions (lbs CO_2)	Climate			
	Cold	Temperate	Hot and Dry	Hot and Humid
Operational	69	59	68	65
Construction	12	16	12	14
Waste	12	16	12	13
Water	7	9	7	8
Total	100	100	100	100

Note: Construction, waste, and water are constant, and transportation is not included. All numbers in percent.

Climate and Architecture

from HEED and Design Builder, most of the emissions from operational energy are produced by space heating, 36% plus an additional 6% for fans and blowers. Design Builder indicates a need for more cooling during many midseason days than HEED.

Emissions from operation are a smaller proportion of all emissions in the temperate climate, averaging 3,857 kg of CO_2e (59% of the total). Heating and cooling make up a smaller portion of the total, and water heating and lighting become more important factors.

Operational energy in the hot and dry climate accounts for 5,807 kg of CO_2e, which is a large portion of total emissions (68%). Most of these emissions (41%) are from cooling, plus 8% for fans and blowers.

Operational energy in the hot and humid climate accounts for 5,052 kg of CO_2e, or 65% of the total building emissions. Most of the emissions from operational energy in a hot and dry climate are from cooling (40%), plus 7% for fans and blowers, which are part of the HVAC system.

In this reference house, operation is the single largest source of emissions (Table 4.2) and the contributor with the highest percentage. Emissions from operation are highest in the cold climate, followed closely by the warm climates, and are lowest in the temperate climate (Figure 4.39). Total emissions are also higher in the cold climate (8768 kg CO_2e/yr), followed by the hot and dry climate (8539) and the hot and humid climate (8148). The lowest value is 6537 in the temperate climate, demonstrating the impact that heating and cooling have on total emissions. Looking into the emissions from operation, emissions from heating and cooling vary by climate but are usually the most important contributor (Figure 4.40). Of course, heating is a larger contributor in a cold climate and cooling in a warm climate.

Operation is the largest source of emissions in all climates and is most significant in the cold climate (69%) and least significant in the temperate climate (45%). The emissions from heating constitute the most significant factor in the cold climate (37%), whereas cooling is the most significant in the hot and dry climate (41%) and in the hot and humid (40%) climate. In the temperate climate, cooling and lights (19%) are more significant than heating and cooling. Appliances are also a significant

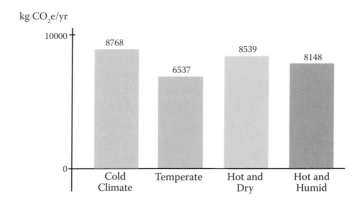

FIGURE 4.39 Effect of climate on emissions from operation (construction, waste, and water are constant).

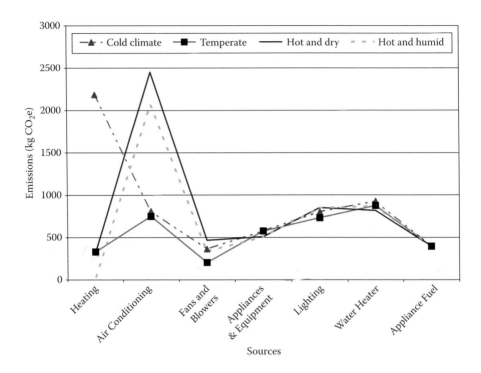

FIGURE 4.40 Effect of climate on source emissions from operational energy (averaged from HEED and DesignBuilder).

percentage of emissions in the temperate climate. For heating, a gas furnace is used; if electricity was used for heating it would have a larger proportion of the total. Also, the role of appliances might be underestimated; data can be modified in the simulations to include more energy use from appliances in our ever more wired homes.

Because operational energy, which includes heating and cooling, has the most impact on emissions, addressing it should be a priority. This can be accomplished with energy efficiency strategies and good old-fashioned passive solar. Photovoltaic and wind systems can also reduce operational emissions by providing clean energy. However, other strategies that affect emissions from construction, water, and waste must also be implemented. These include living close to work, working at home, recycling, and reducing water consumption. Implementing these design strategies (better envelope, more efficient HVAC system, and passive strategies) can dramatically reduce emissions in a typical home in all climates.

5 Solar Geometry

The sun is the source of all life on our planet. It is a huge fusion reactor that works by merging lighter atoms into heavier atoms (hydrogen into helium). This can happen only with the very high temperatures and pressure that are in the nucleus of the sun. By weight, the sun is 70% hydrogen, 28% helium, 0.5% carbon, 0.5% nitrogen, 0.5% oxygen, and 0.5% other elements.

Because the Earth is tilted at a fixed axis of about 23.5 degrees, the northern hemisphere faces the sun in June and the southern hemisphere faces the sun in December. On the northern hemisphere's summer solstice around June 21, the sun's rays will be perpendicular to the earth's surface along the Tropic of Cancer. This is the longest day in the northern hemisphere and the shortest in the southern hemisphere. Around December 21 the situation is reversed, and this is the shortest day of the year in the northern hemisphere and the longest in the southern hemisphere (Figure 5.1).

The temperature of the air and land is mostly a result of the effect of the solar radiation absorbed by the land or the water, which then heats or cools the air above it. As latitude (LAT) increases, winter occurs because the days are shorter with a corresponding reduction of the amount of daylight and solar radiation. Another factor is that in the winter the sun is lower in the sky and its intensity is spread over a larger surface. This is called the cosine effect and is explained graphically in Figure 5.2.

Slopes also affect the amount of radiation falling on a surface. If the angle of the slope is perpendicular to the sun's rays, as in a south-facing slope in the northern hemisphere (A), it will receive more solar radiation than a flat surface (B) (Figure 5.3). A surface facing away from the sun will also receive less solar radiation, which is the case of a north-facing slope in the northern hemisphere. There can be significant temperature differences between a north-facing and a south-facing slope in the same climate. Trees can also block solar radiation and reduce the amount of radiation that is reaching a surface (Figure 5.4).

5.1 THE SUN IN THE SKY VAULT

Several concepts can help in locating the sun in the sky vault: solar declination (DEC), hour angle (HRA), solar altitude (ALT), and solar azimuth (AZI).

5.1.1 Solar Declination and Hour Angle

The Earth's axis of rotation is not perpendicular to the sun/Earth plane; it is tilted 23.5 degrees. This causes the North Pole to tilt toward or away from the sun,

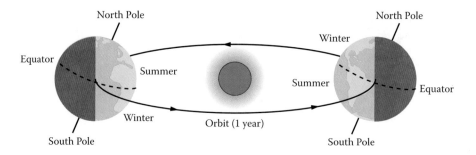

FIGURE 5.1 Earth's rotation around the sun with annual variation.

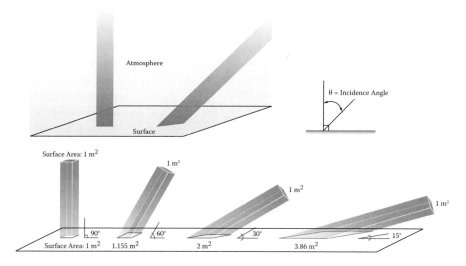

FIGURE 5.2 Effect of the incidence of solar radiation on a surface.

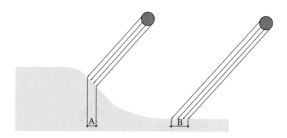

FIGURE 5.3 Effect of the slope on incident solar radiation.

Solar Geometry

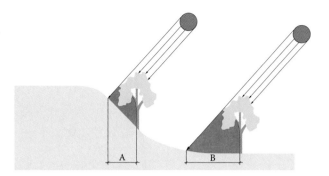

FIGURE 5.4 Effect of trees on solar radiation.

depending on where the Earth is in its orbit around the sun. The solar declination is the angle between the sun's rays and Earth's equatorial plane, or the angle between the Earth–sun vector and the equatorial plane. The total vertical travel between winter and summer is 47 degrees (Figure 5.5).

5.1.2 SOLAR AZIMUTH AND ALTITUDE

The position of the sun in the sky at any given moment can be determined by two values: the solar altitude and the solar azimuth. These are plotted in an imaginary celestial sphere or sky dome in which we are in the center (lococentric view). We regard the radius of this sphere to be infinite, and we can see only 1/2 of the celestial sphere at any time. The boundary between the visible and invisible portions of the celestial sphere is called the horizon. The poles of the horizon—the points directly overhead and underneath—are called the zenith and the nadir, respectively.

The solar altitude is the vertical angle at which the sun's rays strike the Earth that tells us how high the sun is in the sky (Figure 5.7). It is the angle between the sun and the horizon and is a function of the geographic latitude, time of year, and time of day.

The altitude can be determined using

$$\text{ALT} = \arcsin\ (\sin \text{DEC} \times \sin \text{LAT} + \cos \text{DEC} \times \cos \text{LAT} \times \cos \text{HRA}) \quad (5.1)$$

where DEC is declination.

The solar azimuth is the direction of the sun measured in the horizontal plane from the north in a clockwise direction; zero degrees is the north position, 90 degrees is east, 180 is south, and 270 is west (Figure 5.6). In the United States, zero is used for south, −90 for east, and +90 for west (Figure 5.7). The solar azimuth is also called the bearing angle and can be determined using

$$\text{AZI} = \arccos\ [(\cos \text{LAT} \times \sin \text{DEC} - \cos \text{DEC} \times \sin \text{LAT} \times \cos \text{HRA})/\cos \text{ALT}] \quad (5.2)$$

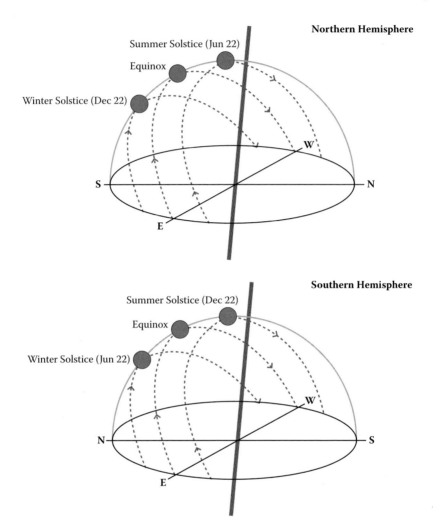

FIGURE 5.5 Variation of solar position in the sky.

5.2 SOLAR CHARTS

The position of the sun in the sky varies during the day and during the year and is a function of the latitude. Complete understanding of solar positioning is necessary for accurate solar design and reasonable climatic response. Solar position can be determined using the previous equations, but it is much simpler and quicker to read sun positions from a table or off a sun path diagram, which can now be generated for any given latitude and longitude.

There are several ways to plot the three-dimensional celestial hemisphere on a two-dimensional surface. The vertical sun path diagram is generated by vertical projections of the sky dome, and the horizontal sun path diagram is generated by horizontal projections from the sky dome (Figure 5.8).

Solar Geometry

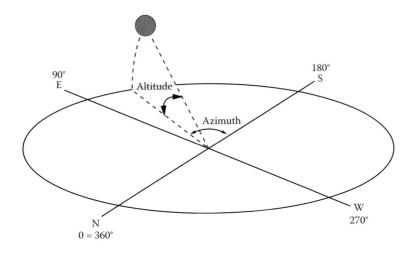

FIGURE 5.6 Solar position angles: altitude and azimuth, in the international system.

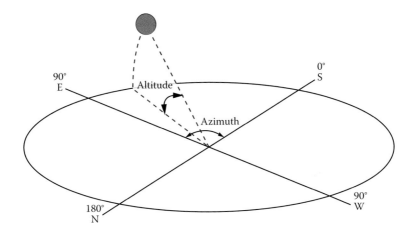

FIGURE 5.7 Solar position angles: altitude and azimuth, in the U.S. system.

5.2.1 Vertical Sun Path Diagram

A vertical sun path diagram uses a cylindrical projection system to project the hemisphere onto a cylindrical vertical surface around it, similar to the earth maps made with a Mercator projection from the Earth's globe. A vertical sun path diagram has the same problem as those maps and provides an accurate representation close to the equator, but it produces a very distorted image closer to the poles, in this case the zenith. The distortion increases because the zenith, which is just a point, is stretched in this map into a line that has the same length as the horizon circle. Figure 5.9 shows a vertical sun path diagram for 34° N latitude..

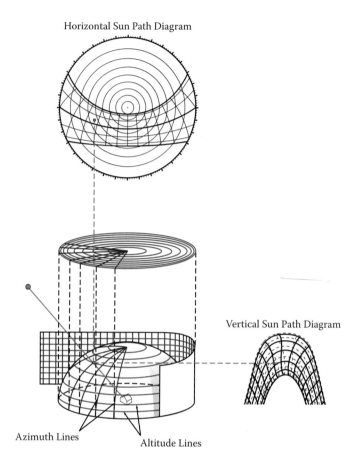

FIGURE 5.8 Projection of the vertical and horizontal sun path diagram.

5.2.2 Horizontal Sun Path Diagram

There are several methods to plot the sky vault in a two-dimensional horizontal sun path diagram. In an equidistant representation system, the altitude is plotted as a set of radial coordinates with evenly spaced altitude circles. In an orthographic projection, the altitude lines are plotted closer together as they approach the horizon because the sky vault is more vertical at lower altitudes and the points project closer to each other, producing distortion toward the horizon. The stereographic projection uses the theoretical nadir point, which is located opposite the zenith, as the center of the projection. Figure 5.10 shows a horizontal sun path diagram for 34° N latitude, and Figure 5.11 schematically shows horizontal sun path diagrams for different latitudes.

It is now possible to use software such as Ecotect to superimpose solar radiation data on a horizontal sun path diagram indicating the amount of solar radiation coming from the sun at specific hours and days (Figure 5.12). This provides additional information that can be used in the design of shading systems.

Solar Geometry

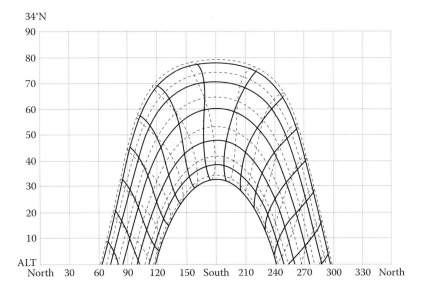

FIGURE 5.9 Vertical sun path diagram for 34° N latitude.

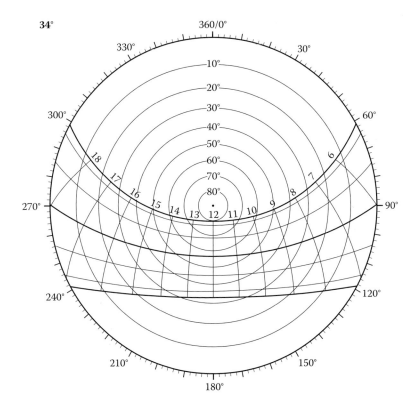

FIGURE 5.10 Horizontal sun path diagram for 34° N latitude.

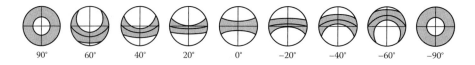

FIGURE 5.11 Schematic horizontal sun path diagrams for different latitudes.

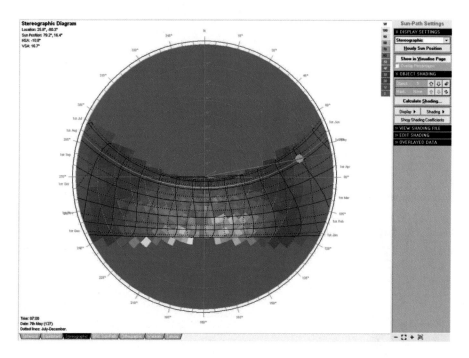

FIGURE 5.12 Horizontal sun path diagram with radiation data generated by Ecotect.

5.3 SHADING THE BUILDING

Depending on the quality of the window, heat flows through them by conduction, convection, and radiation and can be a very important part of heat gains and losses in a building. These processes will be discussed in more detail in Chapter 6; however, in general heat losses in the winter are by conduction, radiation, and infiltration, and heat gains are by radiation. Heat gains and losses in the summer can be by conduction, radiation, or ventilation, depending on the indoor–outdoor temperature relationship. The shading systems should provide shade when it is necessary to block solar radiation (overheated period) and should allow solar radiation when it is necessary to heat the building (underheated period).

It is possible to shade the windows with external or internal shading devices or using the glass pane itself. Each of these has advantages and disadvantages.

An external shading system has the advantage of blocking the solar radiation before it penetrates the building so more of the radiation stays outside, but it has

Solar Geometry

the disadvantage of increased maintenance due to exposure to the climatic elements. Fixed external shading systems are usually classified as horizontal, vertical, or combined (egg crate) and can of course be designed in many different sizes and variations. Solar geometry is used to design these so that they are of appropriate dimension; they should not be so large that they block too much daylight or so small that they do not provide sufficient protection when it is needed.

External shading devices can also be operable permitting to more precisely adjust to different solar positions, during different times of the year, and better responding to the dynamic nature of weather, which does not always correlate with the seasons. The position of these movable devices can also be adjusted as a function of indoor and outdoor temperature instead of only the solar position as the fixed shading devices (Figure 5.13). These devices are constrained only by the designer's imagination and the fact that the device must block solar radiation.

Internal shading systems are usually operable and provide more flexibility to adjust to rapidly changing outdoor conditions. There are also many types of internal

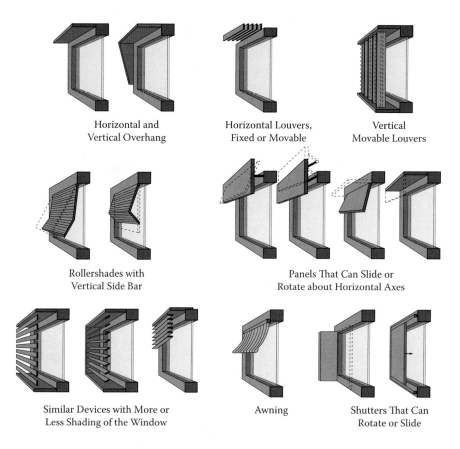

FIGURE 5.13 Examples of external shading devices. (Based on illustrations in Lewis, O., *A Green Vitruvius: Principles and Practice of Sustainable Architectural Design*, 2nd ed., 2001. James and James, London.)

shading systems, which can be horizontal or vertical and include common systems such as horizontal blinds, louvers, shades, and curtains. Horizontal blinds are probably the most effective because they are very adjustable and flexible and can be completely closed to block all solar radiation or can be adjusted to specific angles to permit varying degrees of direct solar radiation or daylight.

Some new high-performance glazing systems have fixed or operable blinds between the panes of glass. Other windows are electrochromic, which means they can be darkened or lightened electronically. A small voltage applied to the windows darkens them, and reversing the voltage lightens them, permitting the amount of light and heat that passes through the windows to be controlled, thus improving their performance compared with a transparent window.

5.4 DESIGN OF THE SHADING SYSTEM

Solar radiation entering through windows can contribute a significant amount of heat into the building, which can be beneficial when heating is required but can also be a liability when cooling is needed. Daylighting, however, should usually be provided during the whole year. A good shading system will provide solar protection during overheated periods and allow solar radiation during underheated periods while providing daylight during the whole year.

Shadow angles describe the sun's position in relation to a building face of a given orientation (ORI) and can be used to describe the performance of the shading device or to specify a device by the type of shadow produced. Two shadow angles help us design an efficient shading system for each latitude and orientation: (1) the horizontal shadow angle (HSA; bearing angle); and (2) the vertical shadow angle (VSA; profile angle).

5.4.1 HORIZONTAL SHADOW ANGLE

The horizontal shadow angle is the difference in azimuth between the sun's position and the orientation of the building face when the edge of the shadow falls on the point considered (Figures 5.14 and 5.15). A smaller HSA angle means a larger fin. The horizontal shadow angle can be calculated using

$$HSA = AZI - ORI \qquad (5.3)$$

HSA is positive when the sun is clockwise from the orientation and negative when it is counterclockwise. When the HSA is between the absolute values of 90 and 270, the sun is behind the elevation.

5.4.2 VERTICAL SHADOW ANGLE

The vertical shadow angle, or profile angle, is the altitude of the sun projected to a plane perpendicular to the building face and measured in this same surface (Figures 5.16 and 5.17). When the sun is parallel to the normal of the surface, then

Solar Geometry

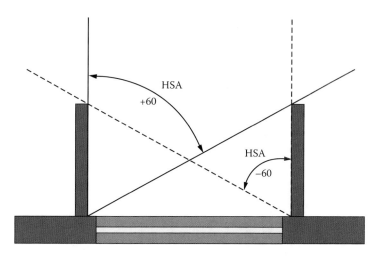

FIGURE 5.14 The horizontal shadow angle in plan view.

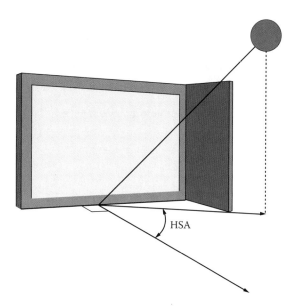

FIGURE 5.15 The horizontal shadow angle in 3-D view.

the solar altitude is equal to the vertical shadow angle. In any other case, when the sun is not perpendicular to the elevation, its altitude angle is projected parallel to the building face onto the perpendicular plane, and the VSA will be larger than the ALT. A lower VSA angle means a larger overhang. The vertical shadow angle can be calculated by

$$\text{VSA} = \arctan(\tan \text{ALT}/\cos \text{HSA}) \qquad (5.4)$$

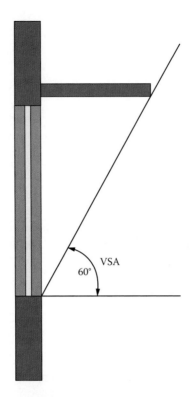

FIGURE 5.16 The vertical shadow angle in section view.

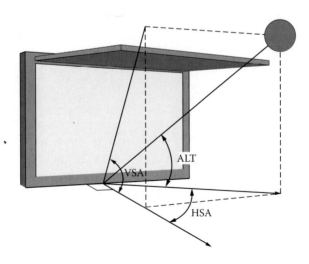

FIGURE 5.17 The vertical shadow angle in 3-D view.

Solar Geometry

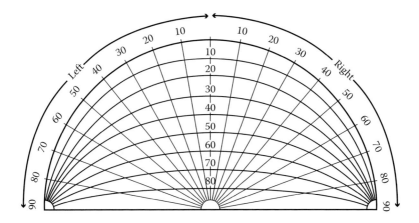

FIGURE 5.18 The shadow angle protractor.

5.4.3 Shadow Angle Protractor

The shadow angle protractor is a semicircular protractor with two sets of lines that are used to calculate the horizontal and vertical shadow angles (Figure 5.18). The protractor has radial lines marked 0 at the center, −90 to the left, and +90 to the right, which give readings of the HAS. It has arcual lines that coincide with the altitude circles along the center line but then deviate and converge at the two corners of the protractor and give readings of the VSA. The shadow angle protractor is used with the sun path diagram to determine the ideal vertical and horizontal shadow angles. It is important that the overhang be designed to the optimum dimension—not too large and not too small. If it is too large, unnecessary material is being used, blocking more daylight and creating darker interior spaces, which might be too cold if heating is needed in the winter. If the shading system is too small, it will not provide sufficient shade during the overheated period, making the space more uncomfortable or requiring more cooling energy.

The following equation can be used to calculate the appropriate dimension for a horizontal element such as an overhang (Figure 5.19):

$$OP = OH/\tan VSA \qquad (5.5)$$

where OP is the horizontal overhang projection of the construction element and OH is the horizontal overhang height—the distance of the shading element to the bottom of the window or the vertical distance that must be covered by the shadow produced by the overhang.

The following equation can be used to calculate the appropriate dimension for a vertical element such as a fin (Figure. 5.20):

$$FP = WW/\tan HSA \qquad (5.6)$$

where FP is the fin projection of the construction element, or the fin size, and WW is window width, or the width of the window that must be protected by the shadow of the fin.

How to Determine the Required Fin Projection:

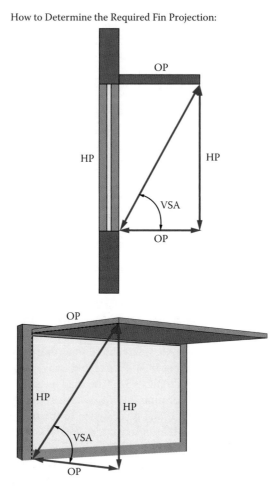

FIGURE 5.19 Calculating a horizontal shading element using VSA.

5.4.4 Example Design Process to Shade a South-Facing Window

To better illustrate the process, consider a designer who must shade a south-facing window (2 m high and 2 m wide) in Los Angeles during the overheated period in a calendar year. The overheated period can be calculated using weather files for the location. For example, Climate Consultant generates a vertical sun path diagram in which the temperatures are plotted for different solar positions permitting to determine the overheated and underheated periods and when shading or solar radiation is needed. Figure 5.21 shows a vertical sun path diagram with corresponding temperatures from June to December in Los Angeles showing that most of the overheated hours are between 9 and 3 p.m. between June and September (Figure 5.22). There are overheated hours outside this area, but it is not cost-effective to design a shading system for all of these hours. These are usually better controlled with operable internal blinds.

Solar Geometry 151

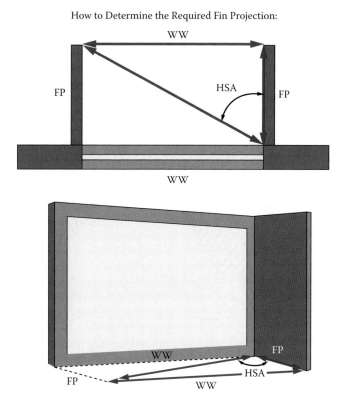

FIGURE 5.20 Calculating a vertical element.

The area that must be protected from solar radiation is now indicated by the shaded area in a horizontal sun path diagram (Figure. 5.23). This area delimits the period during which that building facade must be shaded, as determined by the analysis of climate data and solar geometry. Because it is a south-facing window, a line is drawn from east to west in the sun path diagram to represent this facade, and an arrow perpendicular to this line indicates the orientation of the facade (Figure 5.24). Any point behind this line means that the sun is behind the facade and will not see the sun during those periods. This facade will not receive any sunlight before 8 a.m. or after 4 p.m. in June but gets progressively more radiation until between March and September, when it receives solar radiation during the whole day. Of course only one orientation in one latitude can be done at a time, represented by different sun path diagrams. The advantage is that the whole year can be visualized with only one image.

The next step is to place the solar protractor above the sun path diagram (Figure 5.26) with the bottom of the sun path diagram coincident with the facade line and the radius of the solar protractor matching the radius of the sun path diagram. The goal is to determine the angles (VSA or HAS) that best cover the shaded area. For the south facade it is clear that the VSA is the most effective to protect during the underheated period. If the VSA is too small, it will generate a component that

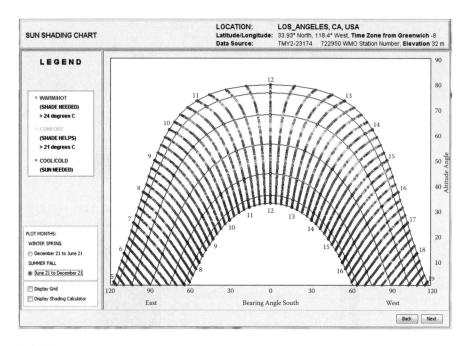

FIGURE 5.21 Vertical sun path diagram with temperatures for Los Angeles (June to December).

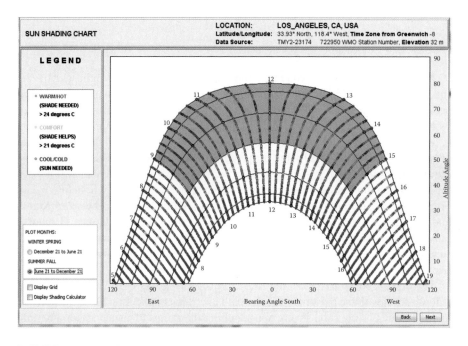

FIGURE 5.22 Vertical sun path diagram with temperatures for Los Angeles (June to December) with overheated areas for which we will design the system.

Solar Geometry

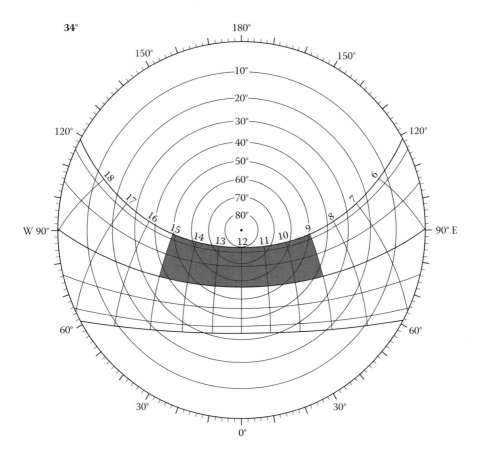

FIGURE 5.23 Horizontal sun path diagram with overheated area as determined with Climate Consultant.

will cover much more than is needed, again wasting material and blocking daylight. If the angle is too large the component will permit unwanted solar radiation that will heat the building. In this example the angle that best covers the shaded area is about 62 degrees, so the desired VSA is 62 degrees. The VSA is drawn on the shadow protractor and then the shaded area of the protractor, called the shadow mask, and in this case indicated in green is the area shaded by a horizontal element that is sized to provide that VSA protection (Figure 5.27).

To calculate the overhang needed to provide this VSA for a window height of 2 m, use Equation (5.5):

$$OP = WH/\tan VSA$$

$$OP = 2 \text{ m}/\tan 62$$

$$OP = 2 \text{ m}/1.88$$

$$OP = 1.06 \text{ m}$$

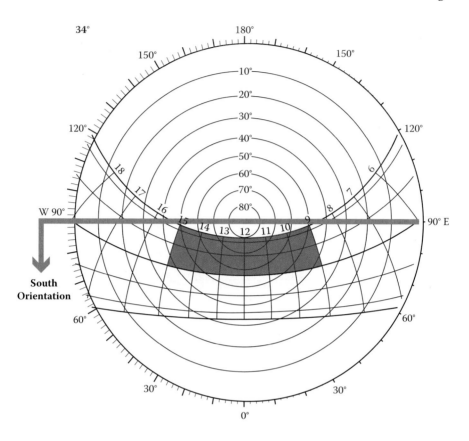

FIGURE 5.24 Overheated area in the sun path diagram with facade orientation.

The advantage of defining the shading system by using the VSA angle is that because the VSA angle is the defining element of the shading system, many different options can be generated from this system: for example, two horizontal elements separated 1 m from each other and half the dimension of the original overhang (0.53 m); or sloped elements that provide the same angle. These can be progressively divided and subdivided until they are very small such as blinds, because the components can be defined as a function of the VSA. This shadow mask shows solar protection achieved by a horizontal element that extends beyond the window. A common error is to think that this level of solar protection can be reached with a horizontal element placed directly above the window and extending neither to the left or the right of the window. If it did not extend beyond the window then it would receive solar radiation from its sides. It is also recommended to place the overhang somewhat above the top of the window to completely expose the window at noontime during the summer.

It is immediately apparent from this example that for this latitude in an equatorial-facing facade, in this case a south facade in the northern hemisphere, an overhang will easily provide the required solar protection—shade in the summer and free heating from the sun in the winter.

Solar Geometry

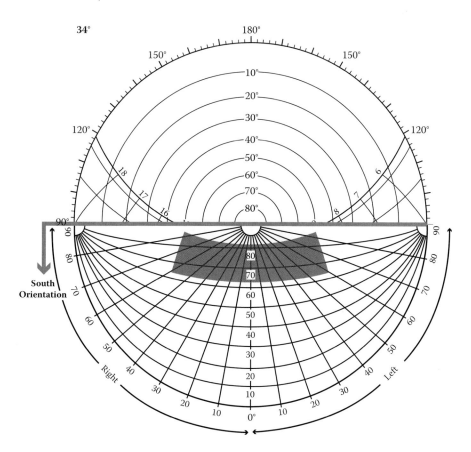

FIGURE 5.25 Sun path diagram with solar protractor.

5.4.5 Example Design of a Shading for a Southeast-Facing Window

To shade the same window, which is now facing southeast, a similar process is followed. The normal orientations of the facade line and the shadow angle protractor are now both facing southeast (Figure 5.27). It is immediately apparent that the facade will not receive solar radiation during many afternoon hours after around 1 p.m. (depending on the month). It is determined that a VSA of about 47 degrees provides the required shade. To calculate the dimension of the overhang, the same equation is applied. An overhang of 1.86 m provides the necessary shade indicated by that overhang.

$$OP = WH/\tan VSA$$

$$OP = 2 \text{ m}/\tan 47 = 1.86 \text{ m}$$

The horizontal shadow angles are determined using the radial lines from the center of the protractor and indicated on either side. A first test is done with the fin

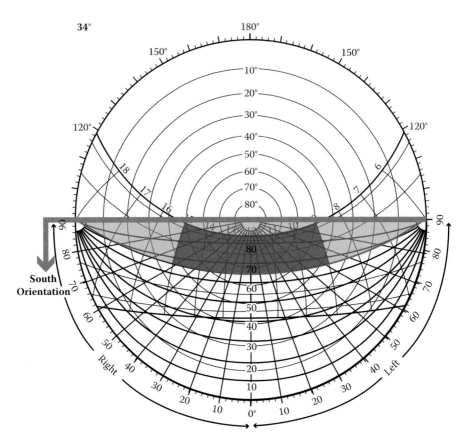

FIGURE 5.26 Solar protractor above the sun path diagram with a shadow mask.

on the north side of the window. An HSA of 30, which would be a significant fin, on the north side of the window (Figure 5.28) does not provide any shade at all. A fin with the same VSA on the south side of that window would provide some shade from around noontime (Figure 5.29); however, this does not include solar radiation that enters when the sun is above the fin. A combination of horizontal and vertical components will properly shade this orientation.

$$FP = WW/\tan HSA$$

$$FP = 2 \text{ m}/\tan 30 = 3.46 \text{ m}$$

Weather files combined with sun path diagrams that also show temperature and solar protractors can be used to design horizontal and vertical shading elements for different orientations.

These methods are useful to size shading systems. However, buildings will have different levels of insulation and internal gains, which will affect thermal

Solar Geometry

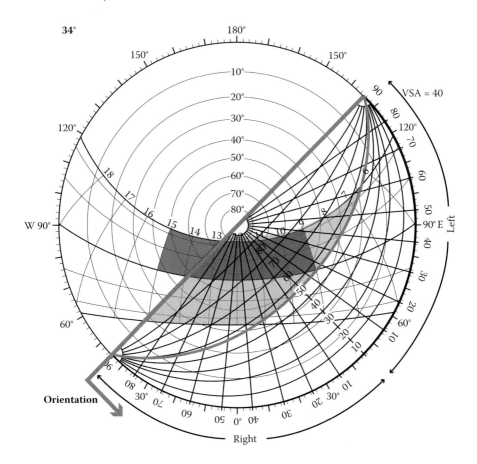

FIGURE 5.27 Southeast facade.

performance. The only way to determine if the shading system is working is to include the shading system in the energy model.

5.5 SUNDIALS

A sundial can be used to analyze shading in a physical model. Sundials are also made for each latitude and require a gnomon of a particular height to cast the proper shadow. The length of the gnomon is indicated in each sundial so that enlargements or reductions can be made (Figure 5.30). To use the sundial it must be pasted on a piece of cardboard in the same orientation as the model that is being tested to indicate the date and time for each shadow option as the model is rotated in different directions. Climate Consultant uses data from weather stations to provide temperature information at different points on the sundial. A good rule of thumb is to provide shade whenever the outdoor temperature is above the lower limit of the comfort zone. For example, if the lower limit of the comfort zone is around 20°C,

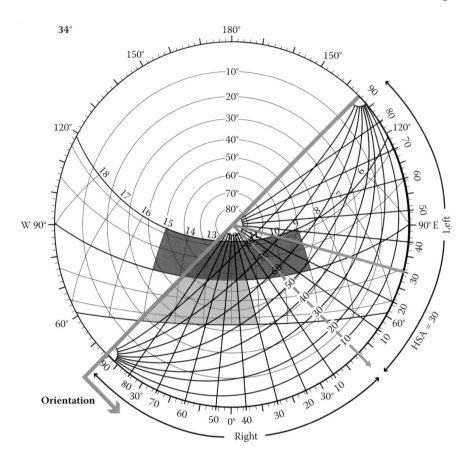

FIGURE 5.28 Solar protractor on the southeast facade.

then shade should be provided whenever outdoor temperature is above this temperature. Thus, if the analysis is being done using the sundial in Figure 5.31, shading should be provided whenever the temperature is above 21°C, the lower portion of the comfort zone.

5.6 SITE ANALYSIS

Sun path diagrams can be used to determine solar potential of a location, especially to determine the placement of photovoltaic systems or passive gains in a solar building. Building shading on outdoor spaces has an important effect on its perception and use. If it is intended for winter use and is shaded by a neighboring building, it will probably be dark, cold, and unused; nor will it be used if it is intended for use during hot weather and it is unshaded. Sun path diagrams and shading masks are used to determine shade from adjacent buildings, trees, or other obstacles. The position

Solar Geometry

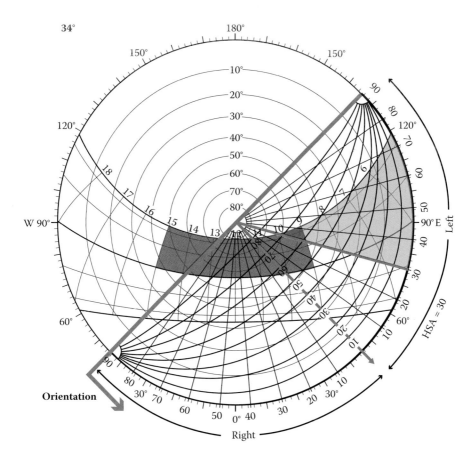

FIGURE 5.29 Shadow mask from a fin in the north side of the southeast window.

of the obstacles is figured by determining the azimuths and altitudes of the edges of the points and plotting them on the sun path diagrams. The area that is intersected by the path of the sun and the enclosed edges indicates when the sun is blocked by obstacles. If the edges are outside the path of the sun, this means that the sun is not blocked by obstacles at any moment. Figures 5.32 and 5.33 show how the same obstacles can be plotted on a vertical and a horizontal sun path diagram. A tree that blocks the sun during a large part of the summer morning, and it is mostly clear during summer close to noontime. Because the position of the obstacles on the sun path diagrams are the result of geometric relationships between the point in the site and the obstacles, the position of the obstacles on the chart changes as we analyze different points. Because they indicate the times during which radiation will be available, these can also be used to establish performance of photovoltaic systems or direct gain systems for passive and active heating. Adjustments in building massing can then be proposed to improve shading or solar access. It is now also possible to generate these sun path diagrams with software such as Ecotect.

160 Carbon-Neutral Architectural Design

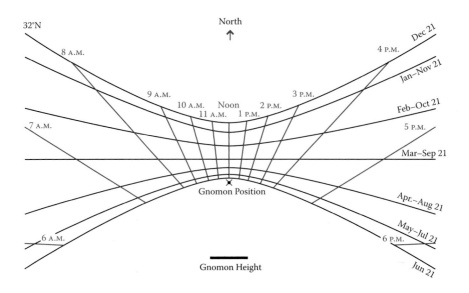

FIGURE 5.30 Sundial for 32° N latitude.

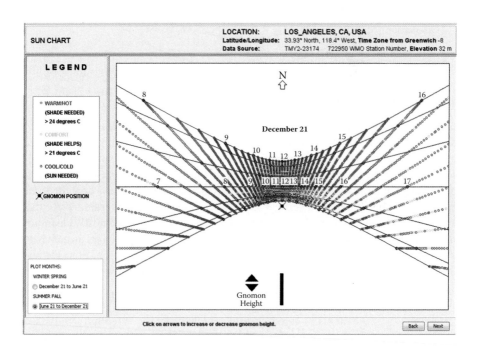

FIGURE 5.31 Sundial with temperatures from Climate Consultant for 32° N latitude.

Solar Geometry

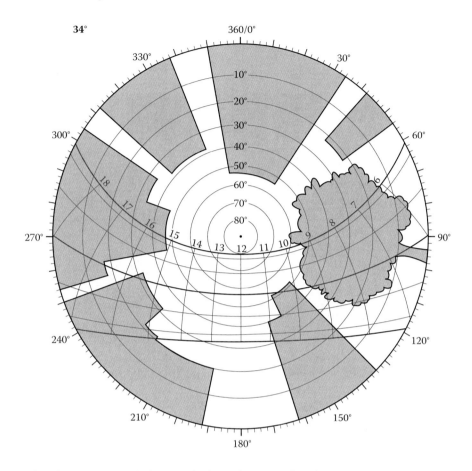

FIGURE 5.32 Site analysis on the horizontal sun path diagram.

In a manual analysis the obstructions must be calculated manually and then be drawn on the sun path diagrams. If the analysis is done digitally, then the obstructions are automatically calculated by the program, adjusting for the relationship between the sun and the obstacles.

5.7 CALCULATING THE IMPACT OF RADIATION ON SURFACES

Building surfaces receive varying amounts of solar radiation depending on the climate and latitude of the site and the orientation and tilt of the surface. Incident solar radiation can significantly impact buildings by affecting the surface temperature of opaque materials and increasing the rate of heat transfer by conduction through walls or by radiation and conduction through windows. Surfaces that receive more solar radiation might require additional shading during the summer or would be suitable for solar hot water collectors or photovoltaic panels or to place a window to provide solar gains to the interior of the building.

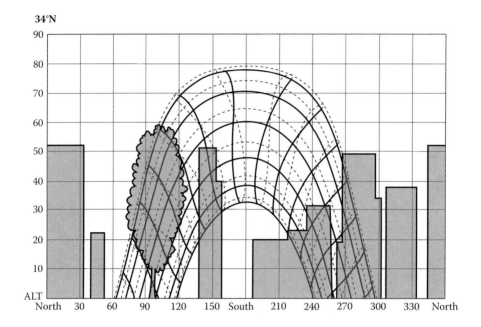

FIGURE 5.33 Site analysis on the vertical sun path diagram.

Using manual methods, it is possible to determine the shading on a building component (e.g., a window), that is, when the building component is or is not receiving direct solar radiation (shaded or unshaded). It does not include the effects of diffuse and reflected solar radiation. In a humid climate a large part of the solar radiation incident on the building could be diffuse, and light-colored surfaces around the building could provide large amounts of reflected radiation. It is now possible to use software in combination with weather files to analyze building components and complex forms with more precision.

With software such as Ecotect and Vasari, the amount of incident radiation on the surface can be calculated so that incident solar radiation on a building's exterior can be compared, allowing designers to make more informed design decisions regarding building geometry. Figure 5.34 shows the analysis of solar radiation (diffuse, direct, and total) on the skin of a competition proposal developed by HMC Architects for the port of Kaohsiung in Taiwan.

5.8 ORIENTATION OF BUILDINGS

The orientation of the building has a considerable effect on its thermal gains. A polar-facing orientation (north facing in the northern hemisphere and south facing in the southern hemisphere) receives direct solar radiation only in the early morning and late afternoon and close to the summer solstice and with vertical elements such as fins.

An equatorial-facing orientation receives more solar radiation during a larger portion of the day as the building is located farther away from the equator, and the

Solar Geometry

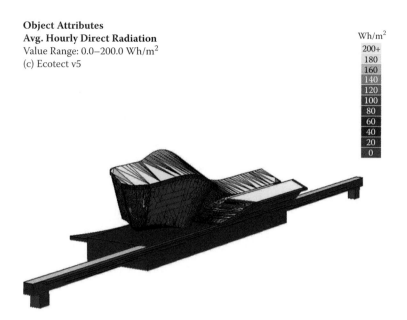

FIGURE 5.34 HMC submittal for the Port of Kaohsiung competition surface analysis of direct radiation. Analysis by E. Santosa. (Courtesy of HMC Archlab.)

angle of incidence also varies according to the latitude, being more perpendicular to the facade as the building is farther north. Equatorial-facing façades are most effective in mid to high latitudes because it is easy to adjust seasonal solar gains with horizontal elements.

In east-facing latitudes in the morning and west-facing latitudes in the afternoon, solar radiation enters the facade at low and difficult-to-protect angles, and to block the sun it is necessary to block views and reduce daylight. This effect is aggravated because of higher afternoon temperatures increasing heat gains by conduction.

In lower latitudes close to the equator, the roofs have the greatest solar gains because they receive most of the solar impacts as the sun attains a high position in the sky very rapidly. In all latitudes the roof gains more in the summer and less in the winter. This is in direct contrast to an equatorial-facing façade, which tends to gain more in winter than in the summer through most of the temperate climates.

6 Heat Exchange through the Building Envelope*

6.1 HEAT TRANSFER THROUGH THE BUILDING ENVELOPE

The design of the enclosure has an important effect on the performance of the building: The envelope affects the thermal interaction with the outdoors, the availability of daylighting, the psychological connection with the outdoors, and many aspects of occupant comfort. This chapter deals with the control of the flow of energy between the interior and the exterior of the building, through the enclosure, to achieve thermal comfort.

Energy is the ability to or capacity for doing work, and heat is a form of energy that appears as molecular motion in substances or as electromagnetic radiation in space. The most common units for heat are the British thermal unit (Btu) and the joule (J). One Btu is the amount of heat required to raise the temperature of 1 pound of water through 1°F (58.5°F–59.5°F) at sea level (30 inches of mercury). The standard unit of energy in the International System of Units (SI) is the joule, equal to the work done when a force of 1 Newton acts through a distance of 1 meter; 1 J is 1 watt-second. The kilowatt-hour (kWh) is used to measure electrical energy and is equal to the work done by 1 kilowatt acting for 1 hour; 1 kWh of electricity is 3.6 megajoules, or 3,412 Btu.

Temperature is a physical property that quantitatively expresses the presence of heat in a substance; the same material (same density and specific heat) with a higher heat content has a higher temperature. The basic unit of temperature in the International System of Units (SI) is the kelvin (K). However, for everyday applications, the Celsius scale is usually used, in which 0°C corresponds to the freezing point of water and 100°C is its boiling point at sea level. In the Inch Pound (IP) system, the Fahrenheit scale is used, in which water freezes at 32°F and boils at 212°F.

A building will always try to attain thermal equilibrium with the outdoors, and heat can flow toward the interior or the exterior but always from a higher to a lower temperature. Heat transfer through the enclosure can be by conduction, radiation, or convection. These flows can occur through the different components of the building fabric: Conduction occurs through the solid components of the building, radiation occurs through materials that have transparency and also affects opaque materials, and convection occurs through any opening or crack in the building. Mechanical systems also move heat between the inside and outside of the building but are not discussed in this book.

---

* Much of this chapter was first developed in La Roche, P., Quirós, C., Bravo, G., Machado, M., and Gonzalez, G. (2001). Keeping Cool Principles to Avoid Overheating in Buildings, Plea Note 6, Passive Low Energy Architecture Assoc. & Research Consulting and Communications.

Understanding these flows helps the designer to identify building design strategies that could be implemented to prevent building overheating or overcooling and to improve the overall thermal quality and energy efficiency of the building. Optimum control of these flows will reduce the energy required for heating and cooling and associated greenhouse gas (GHG) emissions. The ultimate goal is to achieve thermal comfort with strategies that control energy flows through the building fabric without the need for mechanical systems. Implementing these strategies in the building fabric is a necessary first step to improve performance of all systems: passive, active, mechanical heating and cooling, and renewable.

This chapter deals with how to use the building fabric to control flows to reduce summer overheating and winter overcooling. When buildings overheat or overcool, the traditional solution is to pump heat inside or outside; however, these flows can be reduced with appropriate design of the envelope so that it is cooled and heated naturally, using just the envelope.

The different thermal phenomena that must be considered when designing for the control of energy flows are beam or direct solar radiation, diffuse solar radiation, ground-reflected solar radiation, window solar radiation, envelope heat conduction, ground heat conduction, window heat conduction, ventilation heat transfer, electric lights and appliances, heat contribution from occupants, sky longwave radiation, and evaporative cooling (Figure 6.1).

The radiant energy flow from the sun can arrive at ground level in two forms, beam or direct (Gb) and diffuse (Gd), both of which are indicated in Figure 6.1 as (1) and (2). The surfaces around the building also reflect solar radiation (Gr), which can be significant when the albedo of these surfaces is high. Solar radiation can affect heat flow by conduction through opaque surfaces (5) and transparent surfaces (Q_{sw}) (4). The effect of solar radiation on opaque surfaces is included in the heat flow by conduction, and Q_c is the sum of the heat flows by conduction through the solid components of the envelope (opaque or transparent) and includes envelope heat conduction (5), ground heat conduction (6), and window heat conduction (7) in Figure 6.1. Q_{cv} is the result of air mass exchange and is ventilation heat transfer (8) in Figure 6.1. Heat flow (Q_{ir}) is radiant longwave (infrared) cooling, and Q_{ev} is evaporative cooling (11 and 12 in Figure 6.1) and is discussed in the section on passive cooling in Chapter 7.

The heat balance equation, in a simplified form, based on the superposition of steady and periodic thermal phenomenon is used to determine when it is helpful to block or promote thermal flows as shown in Figure 6.1. This equation is adapted from Lavigne, Brejon, and Fernandez (1994) and was used by La Roche, Quirós, Bravo, Machado, and Gonzalez (2001) to organize the strategies that can reduce overheating in buildings.

$$Q_{sw} + Q_i \pm Q_c \pm Q_{cv} - Q_{ir} - Q_{ev} = 0 \qquad (6.1)$$

where
- Q_{sw} = solar heat flow transferred through windows (solar gain)
- Q_i = heat generated inside the building (internal gain)
- Q_c = conduction heat flow through envelope components (includes effect of solar radiation)

Heat Exchange through the Building Envelope

1 Beam (direct) solar radiation (Gb)
2 Diffuse solar radiation (Gd)
3 Ground reflected solar radiation (Gr)
4 Window solar radiation
5 Envelope heat conduction
6 Ground heat conduction
7 Window heat conduction
8 Ventilation heat transfer
9 Electric lights and appliances
10 Occupants heat contribution
11 Sky longwave radiation
12 Evaporative cooling

FIGURE 6.1 Thermal exchanges between a building and the environment.

Q_{cv} = heat flow by ventilation and infiltration
Q_{ir} = net long wave radiation heat flow
Q_{ev} = heat loss by evaporation

In this equation, all heat flows may be instantaneous, but to simplify analysis they are often considered as average values for a period, usually 24 hours. In this case, heat gains and losses represent the daily heating and cooling of the building. Conduction heat flow through the envelope (Q_c) is the sum of the transmittance (U-value) and area (A) of all envelope elements, often referred to as the envelope conductance of the building (Q_c = sum (UA) in units of W/K) and the difference between outdoor and indoor temperatures ($T_o - T_i$). Heat transfer due to air mass exchange (Q_{cv}) is the result of natural or mechanical ventilation or air infiltration: $v_r \times c \times (T_o - T_i)$, which is the product of airflow (v_r = ventilation rate in m³/s) multiplied by the volumetric specific heat of air (J/m³K) often denoted as q_v, or ventilation conductance, and the temperature difference, as presented already. If $T_i > T_o$, then $T_o - T_i$ gives a negative a result (i.e., heat loss); whereas if $T_i < T_o$, then the result is a heat gain. Figure 6.2 illustrates the variables that affect the heat balance in a building.

As a result of solar exposure and indoor heat generation, the interior average temperature of a building without mechanical or passive cooling in the summer will

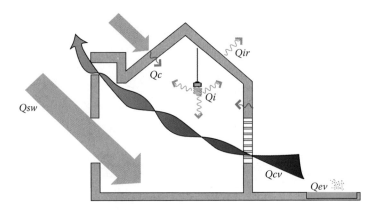

FIGURE 6.2 Variables that affect the heat balance in a building.

always be higher than the outdoor temperature and probably above the comfort zone. Also, because of thermal gains, the indoor average temperature in an unconditioned building in the winter will also be higher than the outdoor average temperature; however, depending on the magnitude of the outdoor temperature it will probably be below the comfort zone. To achieve indoor temperature values closer to the comfort zone, overheating should be reduced in the cooling season, and overcooling should be reduced in the heating season. This is achieved by keeping the heat out in the summer and keeping it inside in the winter.

To better illustrate this process, Tables 6.1 and 6.2, which are both loosely based on Watson's and Labs (1983), as well as diagrams including the components of Equation (6.1) are provided in this chapter to illustrate the strategies that can be used to block, control, or promote heat flows and to avoid overheating or underheating of the building. These same principles are also explained graphically in Figures 6.3 and 6.4. Each heat flow, as well as criteria or design principles that should be applied to reduce its magnitude, is discussed in the following sections.

TABLE 6.1
Strategies to Prevent Overheating

Strategies and Heat Flows	Heat Flows to Be Blocked	Heat Flows to Be Controlled	Heat Flows to Be Promoted
Q_{sw}	■		
Q_i	■		
Q_c	■		
Q_{cv}		■	
Q_{ir}		■	
Q_{ev}			■

Source: Based on D. Watson and K. Labs, *Climatic Design, Energy Efficient Building Principles and Practices*. McGraw Hill, New York, NY. 1983.

Heat Exchange through the Building Envelope

TABLE 6.2
Strategies to Prevent Overcooling

Strategies and Heat Flows	Heat Flows to Be Promoted	Heat Flows to Be Controlled	Heat Flows to Be Blocked
Q_{sw}	■		
Q_i	■		
Q_c			■
Q_{cv}		■	
Q_{ir}			■
Q_{ev}			■

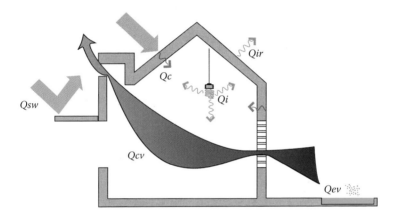

FIGURE 6.3 Strategies to prevent overheating in the summer.

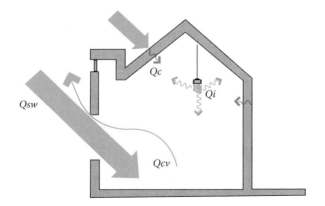

FIGURE 6.4 Strategies to prevent overcooling in the winter.

Considering temperatures and mass and heat flows as average values of a 24-hour period, the average temperature difference between the interior and the exterior of a building (ΔT_m) can be expressed by modifying Equation (6.1) to Equation (6.2). In this equation the numerator includes the flows, and the denominator includes the building components that regulate them:

$$\Delta T_m = T_{om} - T_{im} = \frac{Q_{swm} + Q_{im} - Q_{irm} - Q_{evm}}{q_c + q_v} \quad (6.2)$$

$$T_{im} = T_{om} + \Delta T_m \quad (6.3)$$

where
- q_c = sum ($A \times U$) = envelope conductance (W/K)
- A = surface area of each envelope component (m²)
- U = transmittance of each element (W/m²K)
- q_v = $v_r \times c$ = ventilation conductance (W/K)
- v_r = ventilation rate (m³/s)
- c = volumetric specific heat of air (usually 1,200 J/m³K)
- T_{im} = mean indoor air temperature (°C)
- T_{om} = mean outdoor air temperature (°C)
- q = $q_c + q_v$ = building conductance (W/K)

Indoor building temperature varies around T_{im} with a $\Delta T_{im}/2$ value, where $\Delta T_{im} = T_{i(max)} - T_{i(min)}$; then the interior temperature (T_i) in any given moment can be expressed as

$$T_i = T_{em} + \frac{Q_{svm} + Q_{som} + Q_{im} - Q_{irm} - Q_{evm}}{q_c + q_v} \pm \tfrac{1}{2} sT_i \quad (6.4)$$

where sT_i = swing of indoor temperature from the mean, T_{im}.

Accordingly, maximum and minimum temperatures can be expressed as

$$T_{i\,(max)} = T_{om} + \Delta T_m + 1/2 \Delta T_{im} \quad (6.5)$$

$$T_{i\,(min)} = T_{om} + \Delta T_m - 1/2 \Delta T_{im} \quad (6.6)$$

ΔT_m is calculated from indoor and outdoor temperatures, but ΔT_{im} is the daily variation of indoor temperature.

In warm climates, when cooling is required, ΔT_m should be reduced. The building thermal balance is positive; and because interior temperature is higher than outdoor temperature, $\Delta T_m > 0$. However, in some climates, it is possible to promote heat losses (Q_{ir}, Q_{ev}, Q_{cv}) to provide cooling, which would generate, through natural means, an average interior temperature that is lower than the average exterior temperature: $\Delta T_m < 0$. It is also recommended to reduce internal temperature swing (ΔT_{im}), which can be done by increasing building thermal inertia (thermal mass), and

Heat Exchange through the Building Envelope

solar energy received and transmitted to the interior shading. The building ventilation regime should be adjusted so that ventilation is provided only when it cools the spaces inside.

In cold climates, the sum of all the values in the heat balance equation will tend to be lower than zero and the building thermal balance is negative because interior temperature is lower than outdoor temperature: $\Delta T_m < 0$. In these cases it is necessary to promote heat gains through the windows and if possible through active systems in the walls and roofs (Q_{sw}, Q_c) to provide heating, which would generate, through natural means, an average interior temperature that is higher than the average exterior temperature: $\Delta T_m > 0$.

6.2 HEAT TRANSFER IN BUILDINGS

Heat exists in three different forms: (1) sensible heat (measurable with a thermometer); (2) latent heat (phase change of a material); and (3) radiant heat (a form of electromagnetic radiation).

6.2.1 SENSIBLE HEAT

Sensible heat is energy stored as the random motion of molecules; objects with more random motion have the most sensible heat. Sensible heat can be measured with a thermometer. The heat content of an object (amount of sensible heat) is a function of the temperature, the amount of mass, and the specific heat (C_p). More mass and a higher temperature indicate more sensible heat stored in the material. In Figure 6.5 the top two walls have the same amount of sensible heat because one has three times the temperature and one third the mass. The bottom two walls also have the same amount of sensible heat because one wall has one third the temperature but three times the mass as the other one. Because the heat is a form of energy, the SI unit for heat is J and the IP unit is Btu: 1 kJ = 0.9478 Btu.

6.2.2 LATENT HEAT

Latent heat is the energy required for a change of state of a material. This energy is released or absorbed by a substance during a change of phase (i.e., solid, liquid, or gas), also called a phase transition.

Adding 4.919 kJ to 1 kg of water raises its temperature by 1°C; however, it takes about 2,260 kJ to evaporate 1 kg of water (about 1,000 Btu/lb), which results in no change in temperature. This is because it takes a large amount of energy to break the molecular bonds necessary for a change of state. This heat energy cannot be measured with a thermometer, but it can be used to store and transfer heat, for example, in refrigeration cycles of mechanical equipment.

Latent heat of fusion (melting) is the phase change from solid to a liquid. When ice melts into water, sensible heat is changed into latent heat at a rate of 334 kJ/kg (144 Btu/lb). The ice absorbs this energy to melt; this process is called the latent heat of fusion (Figure 6.6). To change from a liquid to a solid (freeze), the water releases

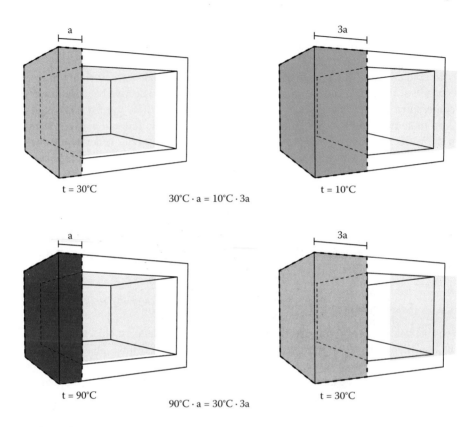

FIGURE 6.5 Relationship of heat content and mass with temperature.

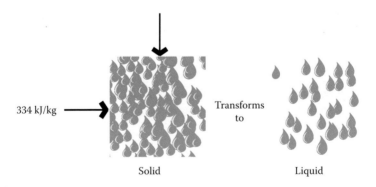

FIGURE 6.6 Phase change from a solid to a liquid.

the same amount of energy that it needs to melt (Figure 6.7). Ice and water can exist in the same glass of water at 0°C (32°F) until all the ice has melted. After this point the water keeps on warming until it reaches room temperature.

Latent heat of evaporation (boiling) is the phase change of a liquid to a gas. When water evaporates, sensible heat is changed into latent heat at a rate of 2,260 kJ/kg

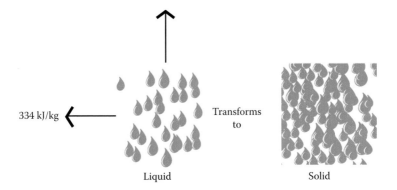

FIGURE 6.7 Phase change from a liquid to a solid.

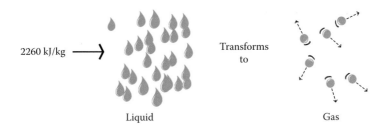

FIGURE 6.8 Phase change from a liquid to a gas.

FIGURE 6.9 Phase change from a gas to a liquid.

(1000 Btu/lb). To evaporate the water it absorbs energy (Figure 6.8), and to condense the vapor to water it releases energy (latent heat of vaporization) (Figure 6.9).

A phase change material (PCM) is capable of storing and releasing large amounts of energy when it changes from solid to liquid and vice versa, and they take advantage of this energy stored and released during the phase change of state of a material to stabilize the air temperature of a space. If the ambient temperature in a space with a PCM rises, the temperature of a solid PCM rises as it absorbs heat around it. When it reaches the phase change temperature it will begin to melt and absorb a large amount of heat at an almost constant temperature. The PCM continues to absorb heat without a significant rise in temperature until all of it is transformed into a liquid. When the ambient temperature around the liquid PCM falls, it releases its stored latent heat as it begins to solidify. A PCM can be designed to change phase at almost any temperature between –5°C and 190°C, and it is very useful in architectural applications if it is designed to change phase close to the comfort zone, reducing the indoor temperature swing, absorbing and releasing energy while it reduces the indoor air temperature swing, and providing the same effect as thermal mass without the bulk. Phase change materials are usually made of a paraffin, fatty acids or salt hydrates. There are many brands that produce different types of PCMs that come in different forms, usually encapsulated in some form, in pouches that can be placed in different locations in the building, or integrated with building materials such as drywall, windows, concrete, and insulation.

6.2.3 Radiant Heat

Radiation is a net transfer of heat from a warmer object to a cooler one by means of electromagnetic waves through space. This transfer of heat occurs in the portion of the electromagnetic spectrum called infrared. All bodies with a temperature above absolute zero radiate energy, so whenever the object faces an air space or a vacuum it will exchange energy with its surrounding surfaces. Hot bodies lose heat by radiation because they emit more energy than they absorb. Radiation is not affected by gravity, and therefore a body can radiate equally in all directions. Thermal radiation can be concentrated in a small spot using mirrors.

6.3 HEAT TRANSFER BY CONDUCTION

Conduction is the process of heat transfer in a solid or a fluid at rest by direct molecular interaction between adjacent molecules. When two objects come into contact, some of their random motion will be transferred from the hotter, more rapidly moving objects (atom or molecule) objects to the cooler, slower objects (Figure 6.10). Conduction occurs only when the molecules are close enough to collide, which is why air is not a good heat conductor and vacuum is not a heat conductor at all. In general, conductivity is related to the density of the material, and materials with higher density are more conductive (Figure 6.11).

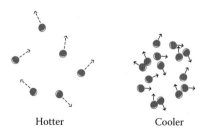

FIGURE 6.10 Heat as a function of random motion.

Heat Exchange through the Building Envelope

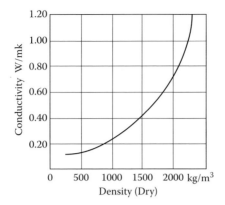

FIGURE 6.11 Relationship between mass and conductivity.

Conduction requires a temperature difference between both sides of the surface, when it is a wall or a building component, or between the solids that are in contact. Heat always flows from a high to low temperature; and if there is no difference in temperature then, there is no flow of energy by conduction (Figure 6.11).

Some characteristics of materials that affect conduction are conductivity, conductance, resistance, and thermal capacity. A concept that defines the overall conductive capacity of a building component composed of several layers is the thermal transmittance or U-value. Two additional factors that define the conductive performance of a material in the building are time lag and decrement factor.

6.3.1 CONDUCTIVITY

Thermal conductivity is a characteristic of the material that indicates its capacity to transmit heat flow. A material with a high conductivity will allow more energy flow. It is expressed in (W/m °C) or (Btu-in/h ft² °F) and is defined as the rate at which heat will flow through a homogeneous material per unit of thickness. In SI units it is defined as the rate at which watts flow through 1 square meter when the temperature difference is 1°C, and in IP units it is the number of Btu/h that flow through 1 square foot of material, 1 in. thick, when the temperature drop through this material is 1°F (under conditions of steady heat flow). As already discussed, materials with higher densities usually have higher conductivities. Table 6.3 shows the conductivity of some materials. More materials can be found in many sources, including ASHRAE's *Handbook of Fundamentals*.

6.3.2 CONDUCTANCE

Whereas conductivity is a property of the material independent of its shape or size, conductance is a property of a body of given thickness. It is the rate of heat flow through a homogeneous material (or combination) of a stated thickness and is measured in W/m² °C or Btu/hr ft² °F. Conductance is useful for materials that come in standard thicknesses.

TABLE 6.3
Conductivities of Some Materials

Good Conductors (W/m °C)	
Copper	280
Steel	52
Aluminum	165
Construction steel	60
Conductors	
Heavy rock	3
Common concrete	2.1
Compressed earth	1
Poor Conductors	
Concrete with light aggregate	0.14–0.27
Glass	0.8
Wood	0.13–0.2
Insulators	
Mineral wool	0.03–0.05
Polystyrene	0.03–0.05

6.3.3 Resistance

The resistance (R) is the reciprocal of conductance and is a much more commonly used value expressed in m² K/W or ft² F° hr/Btu. It is a measure of the resistance of materials and air spaces to the flow of heat by conduction, convection, and radiation, and it measures the insulating value. The lower the R value, the lower the resistance of the material to the flow of heat; for example, R14 means that the material would transfer 1/14 Btu per hour per square foot per °F of temperature difference. All materials offer some resistance to the flow of heat; however, an insulating material will be thousands of times more resistant to the flow of heat than a conductive material, which means that it can be thousands of times thinner.

$$R = b/\lambda \text{ or } R = k/e \tag{6.7}$$

where
  R = resistance
  $\lambda = k$ = conductivity
  $b = e$ = thickness

Air is usually the best resistor for heat flow in a building, but the resistance of the air space is not proportional to its thickness. After about 40 mm, the resistance to

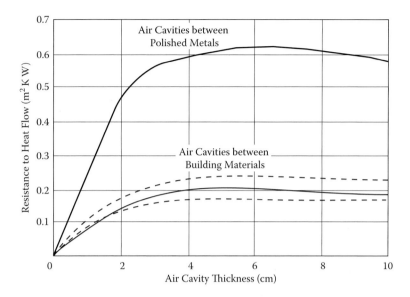

FIGURE 6.12 Resistance of an air space as a function of thickness and surface emissivity.

the heat flow will remain relatively constant; therefore, making an air space larger than 40 mm does not have much effect. The optimal thickness for an air cavity is about 4 cm (a little less than 2 in.). Surfaces that consist of polished materials (low emissivity) facing an airspace provide much better resistance than typical construction materials. If the cavity is larger, then convective currents can occur inside it, increasing heat transfer by convection (Figure 6.12). These convection air currents can be reduced or eliminated by adding lightweight elements inside the cavity that still maintain air in most of the cavity but block convective air currents. For example, a loose tangle of mineral fibers in a wall prevents the air from circulating (Figure 6.13). Thermal resistance of air spaces will also vary depending on the direction of the heat flow.

6.3.4 Thermal Transmittance (U-Value)

The thermal transmittance of an element is the total conductance of a building element and is the inverse of the sum of all the resistances including external and internal air layers.

The U-value is a number that is an expression of the steady-state rate at which heat flows through architectural skin elements. It is measured in W/m² K or Btu/hr ft² °F.

$$U = 1/\Sigma R \tag{6.8}$$

The inner and outer surfaces of a material also offer some resistance to the heat flow (R_{si} and R_{so}, respectively). The reciprocal of these are inner and outer surface conductance (h_i and h_o, respectively) and include convective and radiant components: $h = h_r + h_c$. The degree of surface conductance depends on the position of the surface,

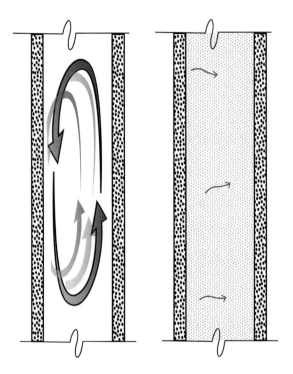

FIGURE 6.13 Heat transfer by convection through an air space blocked by loose material.

the direction of the heat flow, and air movement. Surface resistances should be considered in the calculation of the U-values.

The heat flow through a building's envelope always takes place from warmer surfaces to cooler surfaces and can be from exterior to interior or interior to exterior depending on the temperatures. Therefore, to determine the heat flow, the resistance of the different layers of the wall must be added to the surface resistance, or air-to-air resistance ($R_a - a$). The inverse value of the air-to-air resistance is called the thermal transmittance coefficient (U), and its unit is W/m² °C (Figure 6.14). Thermal transmittance is the value that best defines the amount of heat that can be transmitted through a component of the building, and low U-values block more heat than high U-values.

$$R_a - a = R_{si} + \Sigma R_m + R_{so} \tag{6.9}$$

where R_m is the resistance of material.

$$U = (R_a - a)^{-1} \tag{6.10}$$

6.3.5 Heat Capacity and Specific Heat Capacity

The heat capacity of a system is the amount of heat required to change the temperature of a whole system by one degree. Specific heat capacity is defined as the amount of heat required to raise the temperature of a unit volume or unit mass of a material by one unit of temperature. In the International System of Units (SI), heat capacity

Heat Exchange through the Building Envelope

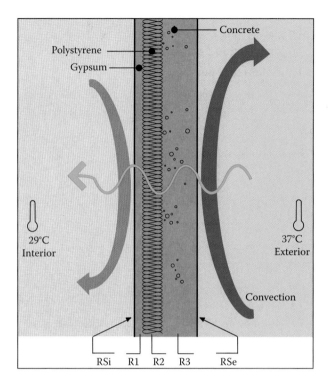

FIGURE 6.14 Thermal transmittance or *U*-value.

is expressed in units of joule(s) (J) per kelvin (K) per unit mass (kJ/kg K). It is also expressed in kJ/kg°C or Btu/lb°F (Table 6.4). It is roughly proportional to the density of the material, with denser materials typically having a higher specific heat capacity than materials with lower density. An important exception is water, which stores much more heat per unit weight than concrete, which is denser. Buildings with massive structures and high heat capacities will act as heat sinks, absorbing and storing heat from the interior of the space (Figure 6.15).

6.3.6 Time Lag

Time lag is the delay in the heat flow measured in units of time (e.g., hours). It is the length of time from the moment when the outdoor reaches the maximum temperature and the indoor reaches the maximum temperature (Figure 6.16). The time lag is affected by the thermal capacity of the material; more massive materials usually generate a larger time lag (Figure 6.17). For example, concrete with a thickness of 30 cm and a density of 1,600 kg/m³ has a time lag of 10 hours.

6.3.7 Decrement Factor

The decrement factor is calculated by dividing the indoor swing by the outdoor swing; thus, a building with a smaller decrement factor has a larger difference between

TABLE 6.4
Thermal Capacity of Several Common Materials per Unit Mass and per Unit Volume

	Per Unit Mass		Per Unit Volume	
	Btu/lb °F	kJ/kg °C	BTU/ft³ °F	kJ/m³ °C
Water	1	4.19	62	4160
Steel	0.12	0.5	59	3960
Stone	0.21	0.88	36	2415
Concrete	0.21	0.88	31	2080
Brick	0.2	0.84	25	1680
Clay soil	0.2	0.84	20	1350
Wood	0.45	1.89	14	940
Mineral wc	0.2	0.84	0.4	27

TABLE 6.5
Resistances and Time Lags of Different Materials

Time Lag	Resistance	Thermal Storage	Materials
Low	Low	High	Metal
High	Low	High	Concrete
Low	High	Low	Styrofoam

these swings. Heavy mass buildings have a smaller decrement factor because they reduce the indoor temperature swing as a function of the thermal mass (Figure 6.18).

$$DF = IS/OS \qquad (6.11)$$

where
 DF = decrement factor
 IS = indoor swing
 OS = outdoor swing

The thermal transmittance of a material is not the only factor that affects thermal behavior. Many combinations of materials in walls can give the same U-value but different mass or even the same amount of mass and with mass location and thermal transmittance in different parts of the wall, which affects thermal behavior (Table 6.5).

6.3.8 HEAT FLOW BY CONDUCTION

In an opaque material, heat flow by conduction takes place directly between adjacent particles of the material when molecules transfer heat from hotter to colder molecules through its higher molecular movement. The rate of heat flow in one homogenous layer varies from one material to another, and, in absence of solar radiation, it is defined by

Heat Exchange through the Building Envelope

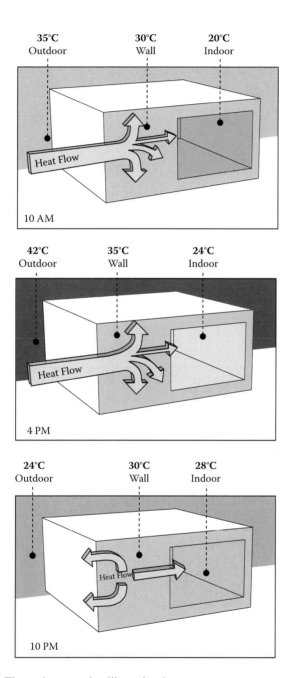

FIGURE 6.15 Thermal mass acting like an insulator.

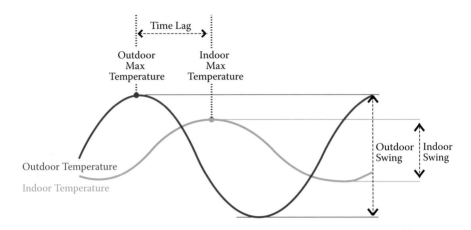

FIGURE 6.16 Time lag.

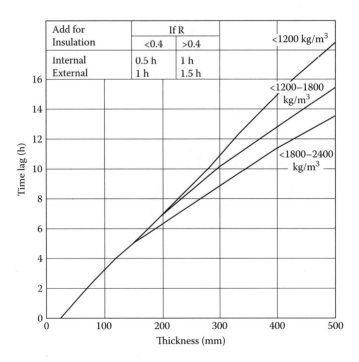

FIGURE 6.17 Time lag as a function of density and thickness.

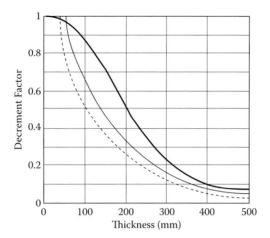

FIGURE 6.18 Decrement factor as a function of density and thickness.

Fourier's equation (6.12). It depends on the thermal conductivity of the material, its width, area, and the temperature difference between both faces of the surface:

$$Q = \lambda \times e^{-1} \times A \times \Delta t \qquad (6.12)$$

where
- Q = heat flow rate or simply flux rate; the total flux in unit of time through a definite surface of material or space, in J/s units, which is the same as watts (W); in IP units, it is Btu/hr
- λ = thermal conductivity of the material (W/m°C)
- e = thickness of the material (m)
- A = enclosure surface or area through which heat can be transmitted; it is perpendicular to the heat flow (m²)
- Δt = temperature difference between the two faces of the surface. In the case of a building's envelope, it is the difference between the external and internal surfaces. If there is no solar radiation, it is assumed that both faces of the surface will have the same temperature as the air (°C)

Fourier's equation can be used to determine heat flow by conduction (Equations 6.12, 6.13). The heat flow is proportional to the thermal transmittance of the material, the area, and the difference in temperature. Appropriate color treatment of exterior surfaces to reduce the incidence of solar radiation can have a positive effect on the reduction of the total heat flow inside the building. This is discussed later in the section on radiation.

The Fourier equation can also be expressed as

$$Q = U \times A \times \Delta t \qquad (6.13)$$

where
- Q = heat flow rate or heat flux (W) (Btu/hr)
- U = thermal transmittance of the element (W/m²K) (Btu/hr ft² °F)

A = area through which heat is transmitted (m²) (ft²)
Δt = temperature difference between the two faces of the surface (°C) (°F)

The application of the design strategies can also be affected by other factors such as the use of the space (daytime, nighttime, or both), air change rate, climatization (natural or mechanical), air humidity, air temperature, and radiation.

In buildings with exterior temperatures close to thermal comfort, insulation is not critical unless radiation affects the building in an important way. In well-ventilated buildings in warm climates with an envelope that is well protected from solar radiation (exterior surface temperature equal to air temperature), thermal insulation will not fulfill any function because indoor temperature could be quite close to outdoor temperature ($\Delta t = 0$). Therefore, heat flow by conduction between the exterior and interior will be very small. In cool climates with outdoor temperatures slightly below the comfort zone, insulation is also not very important because internal gains and heat gains by radiation will usually provide sufficient gains to offset the losses by conduction.

When there is a significant temperature difference between the indoor and the outdoor of the building (at least above 4°C), it is recommended to begin reducing heat flow by conduction. In naturally ventilated buildings, this difference can be the result of daily temperature swings, variations in ventilation flow, or the incidence of diffuse or direct radiation. Insulation is more important in buildings with mechanical heating or cooling, where indoor temperatures are farther away from outdoor temperatures, usually above 6°C in air-conditioned buildings and 10°C in heated buildings and where surface temperature is significantly affected by solar radiation. In these cases, insulation is important to reduce heat gains or losses, which would increase the load on the mechanical cooling equipment.

6.3.8.1 Use of Insulating Material in Walls, Ceilings, and Floors

The first important design principle to reduce heat flow by conduction through the envelope is to use insulating materials with low conductivity and high resistivity. Heat flow is directly proportional to conductivity; a material with a conductivity 50% lower than another material will reduce heat flow by 50%.

Conductivity generally increases with density. Dense materials, such as metals, are more conductive than light materials, such as polystyrene. Air or materials with air have low conductivity, as long as the air is relatively still without major convective air exchanges. The presence of humidity in the material also increases thermal conductivity, because the higher the water content, the larger the amount of molecules in movement. Materials with water in them lose many of their insulating properties.

Some construction materials or building assemblies are made of several components or of several components and air. For example, in a perforated cement or clay brick, the equivalent conductivity is calculated taking into account the proportion of air it contains and the thermal resistance of its material.

In walls, and especially in roofs, combinations of materials that generate structural resistance and insulating capacity must be used. By using layers of insulating materials, good U-values can be obtained with relatively thin widths. It is especially important to control thermal gain in components receiving high solar radiation, such as roofs in low latitudes.

Heat Exchange through the Building Envelope

Zold and Szokolay (1997) classified insulating materials according to how they controlled heat: reflective, resistive, and capacitive insulators. Reflective insulators face an air space and are better at blocking heat transfer by radiation. They must have a low absorptivity and low emissivity (e.g., foil backing on batt insulation). Resistive insulators are those that insulate depending on their width and resistance value. Capacitive insulators act as a function of their width and their density; they have a retarding effect on heat flow and are discussed later.

6.3.8.2   Use of Air Spaces in Walls, Ceilings, and Floors

Density is related to conductivity (Figure 6.11); the materials with the highest densities usually have the highest conductivities. Air, the element with the lowest density and conductivity, is one of the best insulators and is also the cheapest; therefore, it is recommended as one of the layers of a multilayer building assembly, roof, floor, or wall. However, heat flow in these air spaces occurs in several forms: Heat reaches the wall of the air space by conduction and then is transmitted to the other wall by convection and radiation. These air spaces can be ventilated to reduce heat flow by convection, such as in a ventilated attic if the heat flow is downward, otherwise it increases heat flow. To reduce heat flow due to radiation, a reflective or radiant barrier should be installed.

6.3.8.3   Increase the Outer Surface Resistance of Walls and Roofs

The surface of a material also offers resistance to the heat flow, and this superficial resistance depends on its radiative and convective component. The amount of surface resistance depends on the position of the surface, the direction of the heat flow, and the velocity of air. Surface resistance is usually higher in horizontal elements, when the heat flows downward, and when air velocity is lower. It is possible to modify these parameters in the element's surface (e.g., changes in smoothness, texture, inclinations) and thus to increase the resistance of the material, reducing its U-value.

6.3.8.4   Increase Thermal Resistance of Windows

The heat flow through a window exposed to the sun is mainly by radiation and is affected by the angle of incidence of the solar rays and the type of glazing. However, because windows are also objects of solid molecular composition, there is also heat flow by conduction, which can be more or less important depending on the thermal resistance of the window, and the temperature difference between the indoors and the outdoors.

Windows with high resistances are more important when the building is heated or cooled mechanically and there is a larger temperature difference between indoors and outdoors, increasing the heat transfer by conduction. It is important that the glazing system does not block solar radiation in a cold climate, because this radiation can be used for passive heating in the winter.

Currently there are many types of advanced window glazing systems, and their discussion would warrant many pages. Figure 6.19 shows approximate thermal transmittance ranges for different types of windows. To improve thermal resistance, manufacturers usually implement one of these three strategies:

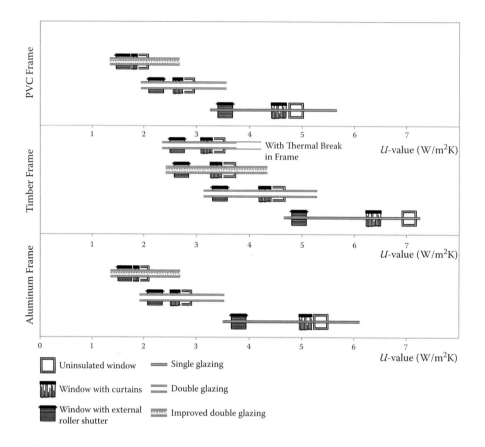

FIGURE 6.19 Thermal transmittance of different window types.

1. Two or more layers of glass with air spaces or with other gases to act as insulators
2. Special coatings that control the amount of solar radiation and thus the amount of heat that is absorbed and transmitted by conduction (and of course by radiation, which is much more significant)
3. Operable elements such as shutters, curtains, or blinds, sometimes inside the window

6.3.8.5 Use of the Thickness of the Architectural Elements as a Regulator of the Building's Indoor Temperature

When the thickness of the material increases, the resistance of the component to the heat flow also increases. However, this increase in thickness has a practical limit due to costs and feasibility of construction. It is much more practical and efficient to use insulating materials rather than regular construction materials to block the heat flow. However, insulating materials have a low density and, therefore, a low structural resistance and should be used in combination with other construction materials that give them rigidity; thus, the need arises for the development of multilayered walls.

Heat Exchange through the Building Envelope

6.3.8.6 Energy Storage Capacity of Materials

Not only does the increase in the material's thickness affect its resistance to the heat flow, as indicated by the very simplified steady-state model, as buildings are not necessarily subject to constant flows of energy; it is also more likely for these flows to vary periodically, generally as a function of the daily 24-hour period, causing a regime of dynamic, nonstatic behavior. On the other hand, all building components have a certain heat storage capacity and can therefore accept energy during the heating process that will be released later during the cooling process, creating a thermal delay and affecting the decrement factor.

The heat storage capacity of a material depends on its mass and therefore on its density and specific heat. The thermal energy stored in a material can be determined with Equation (6.14), where the capacity to store energy is directly proportional to the specific heat of the construction component, its width, and the temperature difference between the material and the surrounding temperature:

$$Q = c \times p \times V \times \Delta t \qquad (6.14)$$

where
  Q = thermal energy stored
  c = specific heat (Wh/kg °C)
  p = density (kg/m³)
  V = volume (m³)
  Δt = temperature difference between the material and surrounding air

Two materials with the same conductivity but different specific heat and density, subject to a constant heat flow, will permit the same heat flow to the interior of the building. However, if the heat flow conditions are periodical, such as in a 24-hour cycle, there can be important differences between the maximum inner values and the moment in which these occur. Therefore, two walls exposed to the same heat flow but with different densities can produce important differences in their indoor swings and thermal delays, even though the average daily heat flow will be the same.

If there is little difference in the maximum and minimum temperature values between day and night, the material will not be able to absorb a lot of energy. Therefore, materials with high thermal capacities are not very useful in areas with thermal swings below 8°C, such as in hot and humid climates, and most of its mass is wasted. However, in locations with higher daily temperature swings, these materials will help to reduce the outer temperature curve through this process of energy absorption and emission. Thus, the swing is reduced. The use of thermal mass as a heat sink is very important in most passive heating and cooling systems and is discussed in the following section.

6.3.8.7 Reduce the Temperature Swing Using Materials with High Density and Thermal Capacity

The delay and reduction of thermal swings inside the space creates temperature differences that can be used to cool or heat up a space. When temperature differences between the indoors and the outdoors are high, with both extremes out of the thermal comfort zone (one above and one below), thermal mass can be used to reduce swing and generate indoor values close to a halfway point that can be closer to the comfort zone.

A material with a high specific heat and density, such as cement, will store much more energy than a light insulating material; therefore, this material will need to absorb much more energy for the same temperature change and will also release much more energy for the same temperature variation. For example, a concrete wall 30 cm wide will need 2 kWh/m² to increase from 10°C to 20°C, whereas the temperature of an insulating material will increase the same number of degrees with only 0.04 KWh/m² (Bansal, Hauser, and Minke, 1994).

It is necessary to know how much mass to use. Most authors use a rule of thumb of about 10 cm (4 in.) of useful thickness of thermal mass for a daily cycle. This useful thickness is the depth of penetration of the heat flow, which refers to the depth that temperature oscillations reach during the processes of energy storage and release and depends on the period; the bigger this is, the larger the thickness of the element. This depth of penetration can be calculated mathematically in many ways, but according to Zold and Szokolay (1997) it can also be calculated with a rule of thumb in which the lowest of the following values is chosen:

1. Half the total thickness of the construction
2. 100 mm
3. The width from the layer in question to the first insulating layer

6.3.8.8 Reducing the Surface Area of the Building

The form factor, sometimes called the skin to volume ratio, is used to determine the energy exchange surfaces. It is the relation between the surfaces of the walls directly in contact with the exterior and the interior volume. The ground surface, walls in contact with another building, or walls buried more than 1 meter deep are usually not included because they are not exchanging energy directly with the outdoors. The form factor can be calculated using

$$F = SE/V \qquad (6.15)$$

where
F = form factor
SE = exterior surface area
V = interior volume

Volumetric loss (G) due to heat transfer through walls is the product of the form factor multiplied by the mean thermal transfer coefficient of the envelope, and represents the insulating level of the building envelope. It is a number that is used in many energy codes to describe the thermal losses through the building skin. To determine this coefficient, losses due to air renovation should be considered; however, only morphological variables are discussed here. Thus, we have

$$G = SE/V \times U \qquad (6.16)$$

$$G = F \times U \qquad (6.17)$$

Heat Exchange through the Building Envelope

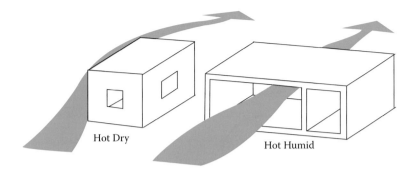

FIGURE 6.20 A compact form is better for a hot and dry climate while an extended form is better in a hot and humid climate.

Greater compactness means less contact with outdoor conditions and in the winter reduced possibilities for heat gain by solar radiation but also reduced opportunities for heat loss. A compact form is appropriate for hot and dry climates, where a compact exterior envelope reduces the amount of incident solar radiation flowing toward the interior of the building.

In buildings that are not mechanically cooled, located in warm and humid locations where ventilation is the most effective mechanism to reduce the psychological effect of high humidity, a compact form is typically not the best solution because extended arrangements provide more opportunities for air movement than compact building forms (Figure 6.20). Once the building is subject to cross-ventilation during daytime hours, its interior temperature tends to follow the exterior pattern, reducing the temperature difference between the interior and exterior and the heat flow through the envelope. Because of this a large surface area does not significantly affect the interior temperature during the day. At sunset and night hours when the winds usually calm down, the bigger area of the envelope allows for faster cooling.

If the interior habitable space is air-conditioned with mechanical equipment and if the exterior surface is exposed to the rays of the sun, the volume should be compact to provide more efficient cooling by reducing surface area exposed to the outside. This is not as important in a super-insulated building or in a milder climate.

In rooms ventilated with passive cooling systems, the volumetric response will depend on the requirements of the adopted system. For systems that store heat, a compact design would be more convenient, but those based on natural convective principles, as night ventilation or passive heating, usually have better performance with a large interior surface more exposed to the indoor air.

In a cold climate, it is even more important to reduce the surface area of the building to minimize conductive heat losses to the outside. Reducing the surface area in contact with the outdoors reduces the amount of energy exchanged to the outdoors, and the most efficient way to do this is to minimize the outside wall and roof areas and thus the surface-to-volume ratio. The sphere is the surface that encloses the most space with the least surface area, and this is probably one of the reasons many early civilizations created dome-like surfaces to protect themselves from the cold elements. Of course, many of these traditional buildings did not have passive solar

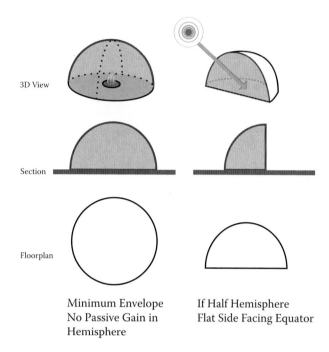

FIGURE 6.21 An ideal form in a cold climate should be compact with the possibility of passive solar gain.

heating systems and instead had a simple central source of heat such a fireplace (Figure 6.21). If passive heating was implemented, it would affect the building form; maybe the skin is expanded where it captured solar radiation for passive solar and is minimized where it did not (Figure 6.21). Also, if the building is very well insulated and the climate is milder, with a smaller difference between the interior and the exterior the form factor becomes a less significant consideration.

6.4 HEAT TRANSFER BY RADIATION

6.4.1 Concepts

Is the process of heat transfer by means of electromagnetic waves. Any two objects that can see each other through a medium that is transparent to light, such as air or vacuum, exchange radiant energy. This exchange can be easily blocked when an opaque object is placed in the path of the rays. Objects at room temperature emit energy in the infrared, and when they are hot enough to glow, they radiate energy in the visible part of the spectrum. Hot bodies lose heat by radiation because they emit more energy than they absorb. The wavelength of the radiation is inversely proportional to the temperature. Higher temperatures have lower wavelengths.

Heat flow by radiation is not affected by gravity; therefore, a body will radiate energy in all directions, is proportional to temperature, is affected by the viewing angle, does not require contact or a medium of transfer, is affected by the surface properties of building materials, is affected by the opacity or transparency of an object, and travels at the speed of light through an empty space.

These radiant exchanges are important because they also affect thermal comfort. Even though these surfaces are not in direct contact with our body, they affect our perception of temperature. One of the indicators of this effect is the mean radiant temperature (MRT), which is the weighted mean temperature of all the objects surrounding the body. It is an average of the temperature of each of the surrounding surfaces, weighted according to the spherical angle and the thermal emissivity of each surface, and is positive if the objects are warmer than the skin and negative when they are colder. MRT is usually measured with a globe thermometer, which consists of a dry-bulb thermometer encased in a 150 mm diameter matte black copper sphere with an absorptivity close to that of the skin. This measurement is useful to determine the effects of radiation on thermal comfort and is an important parameter governing human energy balance because these radiant exchanges affect the temperature of skin, which has a high emissivity and absorptivity of about 0.97.

We can raise the MRT in a cold climate by implementing several strategies: (1) allow the sun to penetrate, which is usually easy to do and free; (2) improve thermal insulation and permit the heating system to warm the surfaces to a higher temperature; (3) install highly reflective surfaces so that they reflect heat back to the body; (4) heat very large surfaces to temperatures a few degrees above the skin temperature; or (5) heat small surfaces to temperatures hundreds of degrees above the skin temperature.

6.4.2 Factors That Affect Solar Radiation

Radiation is affected not by gravity but by the radiative properties of the material with which it interacts, which are affected by the surface of the material. Several surface properties affect radiation: absorptance or absorptivity (a), reflectance or reflectivity (r), emittance or emissivity (e), and transmittance.

Absorptance is a surface quality that indicates the proportion of the radiation absorbed by the material. This radiation is converted into sensible heat within the material, thus raising its temperature (Figure 6.22). Reflectance is a surface quality that indicates the proportion of the radiation that bounces off the material, leaving the temperature of the material unchanged. The reflectance indicates the ability of a surface to reflect solar radiation (Figure 6.22).

Absorptance and reflectance are expressed as decimal ratios as a fraction of the perfect absorber. For example, an absorptivity of 0.74 indicates that 74% of the radiation is absorbed. The sum of the absorptivity and reflectivity of an opaque surface is a unit; whatever is not absorbed is reflected.

Emittance or emissivity is a measure of the ability of a surface to radiate (emit) electromagnetic radiation through the surface, reducing the sensible heat content of the object. It is also compared to a black body, which is the perfect emitter (Figure 6.23). Polished metal surfaces have low emittances, and most other materials have high emittance. Emissivity and absorptivity are not necessarily proportional across different wavelengths. How they relate in a material affects its heat gain by radiation (sensible heat content) and its temperature.

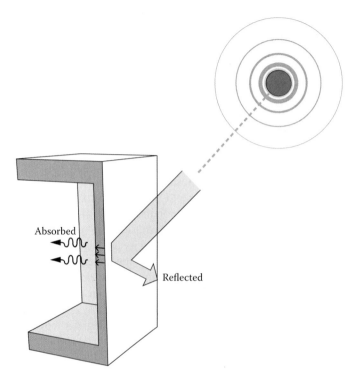

FIGURE 6.22 Absorptance and reflectance.

Most materials have an absorptivity equal to emissivity at the same temperature or wavelength; however, some materials have selective properties and absorb or emit energy differently, depending on the wavelength of the material (Table 6.5). For example, white paint absorbs 20% of the shortwave radiation from the sun and 80% of the longwave radiation from heated objects. Selective surfaces are used in collector panels of solar collectors because they have high absorptivity for solar radiation but low emittance at ordinary temperatures. If we need to lose heat to the sky, we need the opposite effect in which a material will absorb less energy than it emits. Mediterranean lime-whitewashed architecture is a classical example of this effect, and white titanium oxide paints have these properties. Table 6.6 includes the absorptivity and emissivity of various materials.

When the material is not opaque and permits some flow of solar radiation, it has some degree of transparency, or transmittance. The visible transmittance (VT) is an optical property that indicates the amount of visible light transmitted. VT is dimensionless in a scale from zero to one. Different types of glasses have different VT factors, depending on their surfaces, composition, or combination and in windows usually ranges between 0.3 and 0.8. A higher VT means that more light is transmitted (Figure 6.24). In general, glass is transparent to shortwave radiation but opaque to longwave radiation. Shortwave solar radiation from the sun is able to pass easily through a glass, heating objects inside the building, which then emit longwave

Heat Exchange through the Building Envelope

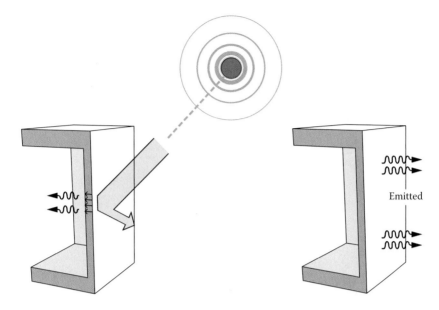

FIGURE 6.23 Emittance.

TABLE 6.5
Absorptivities and Reflectivities of Some Colors

Category	Light	Medium	Dark	Black
Absorbtivity factor	0.5	0.5 < M < 0.7	0.7 < D < 0.9	B > 0.9
Colors	White	Dark red	Brown	Black
	Beige	Light green	Dark green	Vivid blue
	Cream	Orange		Dark blue
		Light red	Light blue	Dark brown

TABLE 6.6
Shortwave Absorptivity and Longwave Emissivity of Some Materials

Substance	Shortwave Absorptivity	Longwave Emissivity
White plaster	0.07	0.91
Aluminum oil paint	0.45	0.9
Red bricks	0.55	0.92
Concrete	0.6	0.88
Gray paint	0.75	0.95
Black gloss paint	0.9	0.9
Perfect black body	1	1

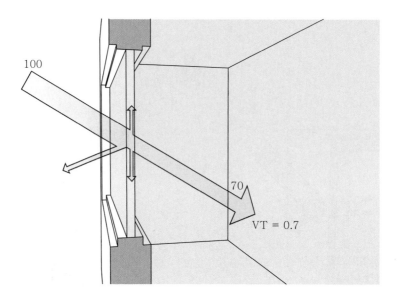

FIGURE 6.24 Transmittance through a transparent material.

radiation that cannot pass through the glass to the outside. This additional heat that is trapped inside the space, warming it, is called the greenhouse effect.

6.4.3 Effects of Solar Radiation

When the external hot surface of an element is exposed to solar radiation, its surface temperature can increase considerably, escalating heat flow by conduction to the interior or the building (because of the larger temperature difference). Solar radiation affects the surface temperature of the element and therefore heat flow by conduction. In warm climates it is necessary to create shade, to use materials with low absorptivity, and to increase surface resistance with different surface treatments. In cold climates it is necessary to promote solar gains and use dark colors. Of course, it is not feasible to change the color of a material as seasons change.

Total shortwave radiation (I_t), which reaches any surface on the earth, is the sum of direct solar radiation coming from the solid angle of the solar disc (I_d), diffuse celestial radiation (I_d), and the reflected radiation of the adjacent surfaces (I_r) (Figures 6.25 and 6.26). The intensity of the direct component on a vertical surface is the product of the normal direct irradiation (I_{DV}) and the cosine of the incident angle (f) divided by the incident radiation and the normal to the surface.

$$I_t = I_{dr} + I_d + I_r \qquad (6.18)$$

$$I_{DV} = I_{DN} \times \cos f \qquad (6.19)$$

$$I_{TV} = I_{DN} \times \cos f + I_d + I_r \qquad (6.20)$$

Heat can be gained or lost by conduction through the envelope; however, in a warm climate solar gains are by far the most important source of heat gains in

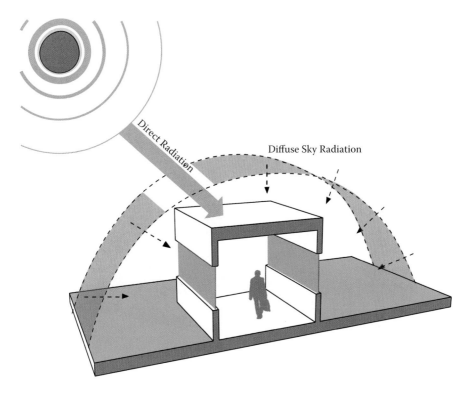

FIGURE 6.25 Diffuse and direct component of radiation received by a building.

a building and in a cold climate can be the only source of external gains. In a warm climate it is necessary to minimize the amount of solar radiation falling on a building envelope. Any solar radiation will be absorbed by the building components, which will be heated and thus will increase conductive gains to the interior or heating the building directly by radiation through the windows. In a cold climate, solar radiation incident on the walls and through the windows will be beneficial because it adds heat to the building, potentially outweighing the losses by conduction due to the temperature differences between the interior and the exterior of the building. Heat gains through the windows are discussed in Chapter 7 on passive heating. The effect of solar radiation is different through the opaque and transparent elements of the building skin.

6.4.4 Opaque Components

The total solar radiation (I_t) incident on the building's walls or roofs is the sum of the direct, diffuse, and reflected components. After it hits the opaque components, part of the incident solar energy is reflected and another part is absorbed by the wall. The exterior side of the enclosure increases its temperature to a level higher than air temperature; part of the energy absorbed is dissipated to the exterior by convection and longwave radiation, and another part is transferred toward the interior by conduction through the wall or roof.

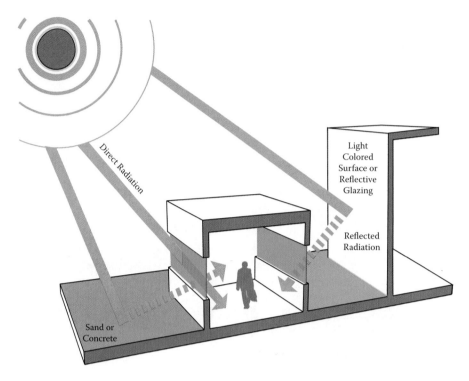

FIGURE 6.26 Radiation can be reflected to the building from surrounding areas.

The effect of this heating by solar radiation on external surfaces is measured by the sol–air temperature, which combines the heating effect of the incident radiation falling upon a building with the effect of the hot air in contact with this surface. The sol–air temperature is the new surface temperature used to calculate heat flow by conduction, accounting for the thermal effect of the incident solar radiation.

$$T_{SA} = T_o + \frac{a \times I_t}{h_o} = LWR \qquad (6.21)$$

If
$$R_{Se} = 1/h_o \qquad (6.22)$$

$$T_{SA} = T_o + a \times I_t \times R_{Se} - LWR \qquad (6.23)$$

where
- T_{SA} = sol–air temperature (°C)
- T_o = outdoor air temperature (°C)
- a = absorptivity of external surface (%)
- I = intensity of solar radiation on the surface (W/m²)
- R_{Se} = outdoor wall surface resistance coefficient (m² °C/W)
- h_o = overall external surface coefficient (radiant and convective) from the exterior surface to the environment. For design purposes, 20 W/m² °C with a wind speed of 3.5 m/s can be used.

Heat Exchange through the Building Envelope

LWR = correction value resulting from the difference between environmental temperature and the temperature of the sky, due to radiant losses from the surfaces of the building. It results in a drop in temperature due to long-wave radiation losses to the sky. For the roof it is about 6°C in a hot and dry climate with a clear sky, 4°C in a hot and humid climate with a clear sky, and about 0°C under cloudy skies. The value for vertical surfaces is negligible because these usually face surfaces with similar temperatures.

Applying this equation makes it clear that in the presence of solar radiation, materials with higher absorptivities will have a higher surface temperature than materials with lower absorptivities and more heat will be transferred to the interior. If a material is shaded from solar radiation, then its temperature will be equal to the temperature of the air.

6.4.4.1 Use of Shading Devices

As was seen in the previous section, solar radiation on external surfaces can increase surface temperatures and thus the heat flow toward the interior of the building. Shading, by providing opaque devices between the sun and the enclosures, can significantly reduce the heat flow to the interior. If the incident radiation (I) is equal or tends to zero, then ($I \times a \times R_{Se}$) will tend to zero too, and the external surface temperature will be close to the temperature of the air (Figure 6.27). The heat flow in the interior of the building (Q) will then be lower than that of a similar portion of the facade that is exposed to the sun.

The conductive heat flow through an opaque element receiving solar radiation in its external face can be expressed as follows:

$$Q = U \times A \times [(T_{SA} = T_o + a \times I/h_o - LWR) - T_i] \qquad (6.24)$$

or

$$Q = U \times A \times (T_{SA} - T_i) \qquad (6.25)$$

where
- Q = conductive heat flux (W)
- U = thermal transmittance (W/°C m²)
- A = area of the wall (m²)
- T_i = indoor temperature

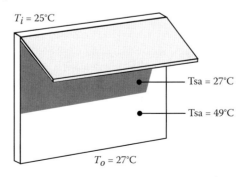

FIGURE 6.27 Solar protection for external walls.

6.4.4.2 Types of Solar Protection

Solar protection is usually classified as horizontal or vertical; however, in this case we will use the convention proposed by La Roche, Quirós, Bravo, Machado and Gonzalez (2001) in which the elements are classified according to how they are placed in relation to the building component. They can be placed parallel to the façade or perpendicular to the façade. Thus, a component that is perpendicular to a flat roof will be vertical, whereas a component that is parallel to a flat roof will be horizontal. These elements can be designed using the concepts of solar geometry explained in previous chapters.

6.4.4.2.1 Parallel Shading Devices

The direct solar component can be blocked with opaque components parallel to the building surface, with an external air circulation space. This system is used mostly in roofs or in east and west façades that require extensive solar protection. It has the disadvantage that it blocks all views from the building component toward the outside, and when used in roofs also blocks solar radiation flows toward the night sky. The energy flow through the opaque component and toward the enclosure can be described as follows:

1. The external face of the sunshade absorbs part of the solar radiation and the rest is reflected.
2. Absorption of the flow dissipates to the exterior by convection and long-wave radiation. Another portion dissipates by conduction toward the interior surface of the sunshade.
3. A portion of the heat flowing through the sunshade dissipates in the air space by convection, and another portion flows toward the external surface of the wall or roof by longwave radiation.
4. Part of the flux absorbed by the external face of the wall or roof dissipates in the air space by convection. Another part is absorbed by conduction to the internal face of the enclosure.

A solar protection for a roof made of prefabricated panels (i.e., concrete, clay ceiling fillings) or galvanized iron sheets or fiber cement blocks incident direct solar radiation but does not allow radiation to escape toward the night sky (Figure 6.28). If the shading system can pivot, then it can be rotated so that it opens at night and the building components lose radiation toward the night sky (Figure 6.29). The Spanish *toldo* is an example of a movable shade that can be opened at night.

An alternate method is using a trellis covered with vines. Evaporation from the surface of the leaves will reduce their temperature below daytime air temperature (Figure 6.30). A green roof provides a similar effect, depending on the type of the vegetation planted on it.

6.4.4.2.2 Perpendicular Solar Protection

Perpendicular solar protection systems are mostly used in walls and can be horizontal or vertical. In mid latitudes, horizontal systems are usually recommended for

Heat Exchange through the Building Envelope

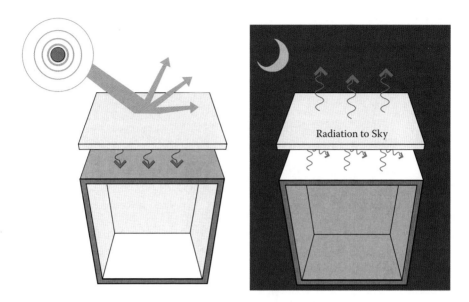

FIGURE 6.28 Shading of a roof with a parallel solar protection system.

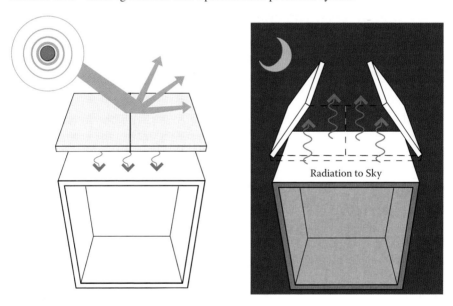

FIGURE 6.29 Operable parallel shade.

equatorial-facing elevations, vertical components for polar-facing orientations, and combinations of these for east- and west-facing orientations.

In addition to blocking direct solar gains, solar protection systems perpendicular to opaque enclosures also provide some heat dispersion with fins. However, if they are connected to the internal structure, they can also create thermal bridges. Parallel

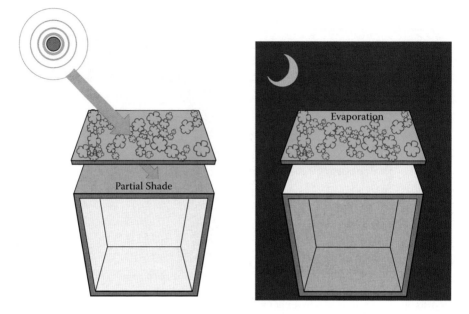

FIGURE 6.30 Parallel shading with vegetation.

solar protection can also provide some heat exchange through convective processes by pressure differences generated in the surface of the envelope due to small temperature variations.

6.4.4.3 Building Volume

The building volume can be conceived to maximize shading when necessary in the cooling season or to maximize solar gains when necessary in the heating season. Buildings can be designed as convex or concave configurations, which, depending on latitude, azimuth, and angle, will either block or receive solar radiation. There is an opportunity here to generate algorithms or generative design systems that provide optimum building skin configurations for different types of climates. A tool in this direction was developed by da Veiga and La Roche (2003). This tool evaluated solar radiation on building envelopes using Maya, a three-dimensional (3-D) modeling and animation software and analyzed the surfaces of complex envelopes with different orientations using the potential amount of solar radiation that each surface section received as a function of the angle of each one with the sun (Figure 6.31). The different surfaces could be analyzed for a particular day and latitude and a specific outdoor average temperature. A fitness function determined and rated the individual and overall performance of the proposed building envelope. This tool worked as a rating system for complex architectural forms to aid in the evaluation of complex building envelopes as a function of incident energy, allowing the performance of each proposed envelope to be determined. The tool was not developed further, but now several tools such as Ecotect incorporate this function, permitting the user to determine optimum orientations for solar hot water collectors, photovoltaic panels, or windows to collect solar gains to the interior of the building.

Heat Exchange through the Building Envelope

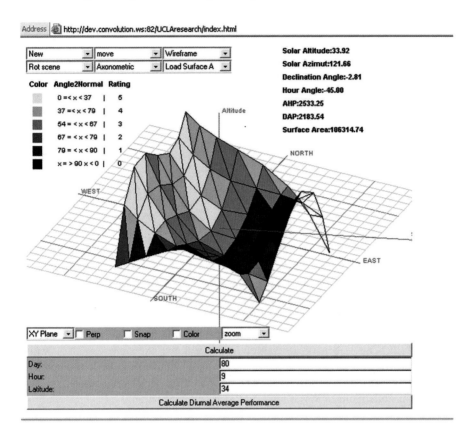

FIGURE 6.31 A computer tool for the analysis of direct solar radiation in complex architectural envelopes: EvSurf.

6.4.4.4 Opaque Surface Finish

Roughness is related to the texture of the building's enclosure surfaces on a small scale. Grades of roughness are established according to the size of the grain, expressed in millimeters. Larger or smaller roughness on the exterior skin of the building has little effect on temperature. However, a very bumpy skin favors radiant exchange between the radiant surface and the sky, but can also reduce connective exchanges.

Highly textured walls have one part of their surface in the shade. The absorption area of the textured surface is smaller than its emission area and will thus be cooler than a flat surface. The enlarged area can also generate an increase in the heat convective transfer coefficient, which will allow faster cooling of the building during nighttime when the surrounding temperature is lower than the temperature of the building.

6.4.4.5 Selection of Absorptive, Reflective, and Emissive Materials for Exterior Surfaces

These concepts were discussed earlier in the chapter. If absorptivity decreases, then the outdoor temperature will be closer to that of the air temperature, and the heat flow toward the interior of the building (Q) in the summer will be lower. The external surface color has a significant effect on the impact of the sun upon the building and the temperature in the interior spaces. Materials with darker colors, such as black, are the most absorptive, whereas light colors are the least absorptive. Radiation delivered by a material from its surface is equal to absorption when it is at identical wavelength. But the solar radiation absorbed by a surface is not necessarily equal to the radiation emitted, because the temperature of this material is much lower than that of the sun.

Some cold materials absorb a reduced proportion of solar radiation but deliver radiation freely. They maintain relatively low temperatures even when exposed to the sun. Other materials absorb a high proportion of solar radiation but have reduced emissivity. These surfaces maintain high temperatures and reduced heat losses. Figure 6.32 shows a selection of materials with different combinations of absorptivities and emissivities that can be used for different purposes.

6.4.4.6 Building Components That Are Transparent to Solar Radiation

Transparent and translucent building components such as windows or skylights have several functions in any climate; they provide daylight, visual contact with the outside, solar gains in the winter, and ventilation in the summer. Windows are major contributors of heat by transmission of radiation and re-radiation toward the interior of the space. This is beneficial in the winter but not in the summer. Glass is also responsible for the greenhouse effect, which is important to consider when calculating thermal gains through windows. In addition to heat gains by radiation, a window can also provide gains or losses by conduction and ventilation.

The following equation is used to determine the heat flow through a glazed opening:

$$Q = A \times I \times f \tag{6.26}$$

where
 A = window area in m^2
 I = solar radiation intensity on the window surface in W/m^2
 f = window glass solar gain factor, which depends on the type of glass and incidence angle, measured from the normal to the surface. It is expressed as a decimal fraction of the incident radiation that reaches the interior and in the United States it is called the solar heat gain coefficient (SHGC)

The value of f varies as a function of the geometrical relationship between the window and the solar radiation, which is a function of the latitude, time of day, and orientation and angle of the component. This radiation has three components: direct, beam, and diffuse. The sum of them is the total radiation upon the surface.

6.4.4.7 Appropriate Window Selection to Control Solar Radiation

Of the radiation that hits the window, part of it is absorbed, part of it is reflected, and part of it is transmitted toward the interior of the space. Because heat gains through

Heat Exchange through the Building Envelope 203

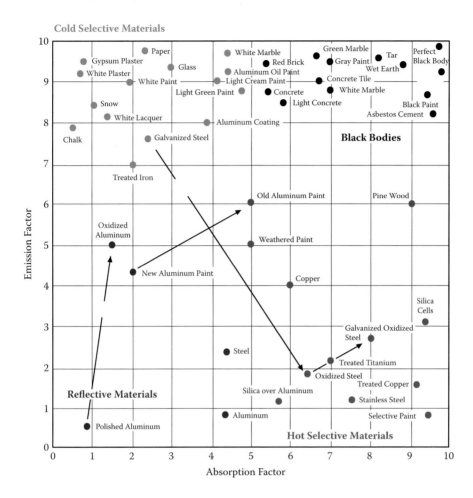

FIGURE 6.32 Emissivity and absorptivity of some materials.

windows and skylights can be an important part of the thermal gains in a building, it is important to select appropriate window types for each climate and orientation.

Because of the need to control this solar radiation, the glazing itself becomes an important mediator between the building and the exterior. Until around the 1950s, most architectural glazing was clear, transmitting 80–90% of visible and total solar radiation, and consisted of a single layer of glass. Now, many types of higher-performance glazing systems regulate energy (heat and light) in different ways.

In a clear single-pane window, most of the solar radiation is transmitted toward the interior of the space (Figure 6.33). In a heat-absorbing tinted glass, a large portion is absorbed, of which most is reradiated toward the interior of the space (Figure 6.34). In a reflective glass, a larger portion is reflected toward the exterior (Figure 6.35). Of course, these numbers are approximations and vary depending on the manufacturer and specific windows.

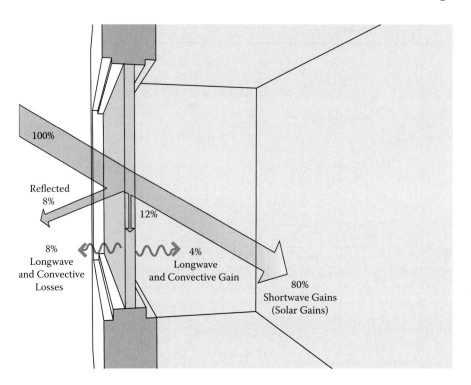

FIGURE 6.33 Solar energy balance in a single-pane window.

The SHGC measures how well a window blocks heat from sunlight and indicates the percentage of solar radiation incident upon a glazed aperture that ends up as heat inside the space. The SHGC does not have a dimension; the maximum theoretical resistance is zero, and the minimum one is calculated based on the performance of the whole glazing unit, measured using the rough opening and not just the glass portion. In reality it ranges from around 0.2 to 0.9 because no window completely blocks the heat or allows all of the flow of heat inside the space. The SHGC is affected by the type of glass, number of panes, tinting, reflective coatings, and shading by the window or skylight frame. In general, if solar gains are required it is preferable to have a high SHGC (allows more heat), and in cooling climates it is preferable to have a low SHGC (blocks more heat). Summer gains can be controlled by shading devices.

SHGC quantifies the total amount of the solar radiation spectrum penetrating through the window, whereas VT quantifies only the visible portion of the spectrum. The relationship between SHGC and VT is expressed as the light-to-solar-gain ratio (LSG) obtained by dividing the VT by SHGC. All windows should provide daylighting. However, windows in hot climates should have a low SHGC because any solar gain through window increases cooling load, whereas windows in cold climates should have a higher SHGC because any gain through windows could reduce heating loads (Figures 6.36 and 6.37). They should all be good insulators

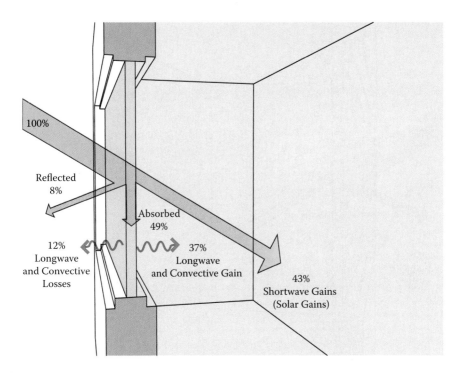

FIGURE 6.34 Solar energy balance through heat-absorbing tinted glass.

and should be airtight to reduce infiltration when they are closed. These two factors become more important as the difference between the indoor and the outdoor temperature increases.

Low-emittance coatings (low-E) are applied to one of the glass surfaces facing the air gap between the layers of glazing. They are microscopically thin, almost invisible metal or metallic oxide layers deposited on a window or skylight glazing surface. The principal mechanism of heat transfer in multilayer glazing is thermal radiation from a warm pane of glass to a cooler pane. A low-E coating blocks a large portion of the heat transfer by radiation through the glazing panes and increases the overall resistance of the window (lower overall U-value). Low-E coatings are transparent to visible light but reduce ultraviolet (UV) transmission through the window. Different types of low-E coatings have been designed to allow for high solar gain, moderate solar gain, or low solar gain. Higher solar gain windows perform better in winter, and lower solar-gain windows perform better in the summer. In cooling-dominated climates, windows with lower solar gain are better because all heat entering the building contributes to increase the heat gain. Depending on the position of the coating, the windows will block the heat entering or leaving the space. These windows must then be selected accordingly for summer or winter use. They must block heat gains in the summer and heat losses in the winter while providing the possibility for gains in the winter.

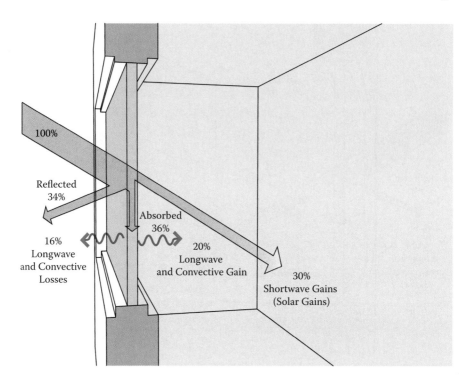

FIGURE 6.35 Solar energy balance in a window with reflective glass.

Several types of photosensitive or photochromatic glass have been recently developed to turn dark when exposed to intense light and to recover their transparency when the luminous source disappears. The transmittance of these windows can vary by about 30% at a time and between 74 and 1%, for example, 1%–20%, 15%–40%, 35%–55%, and 50%–75%.

Windows with U-values below 1.5 W/m²K can be constructed in several ways. Most of them are double- and triple-glazed systems that use coatings with low emissivity and noble gases between the glazing layers. Double-glazed units with low-E coating and argon can be built with values about 1.3 W/m²K, and can achieve up to 0.56 W/m²K. The problem with many of these gas-filled glazing systems is that they lose most of the gas over time.

New windows use silica aerogel, which is transparent to visible solar radiation but is very insulating with a very low thermal conductivity around 0.021 W/m²K, greater than that of still air (0.026 W/m²K). Some of these have been built inside evacuated aerogel glazing constructed as a sandwich by inserting a 20-mm-thick aerogel disk between the two glass panes with U-values around 0.5 W/m²K. Manufacturers such as Kalwall now make systems using Nanogel® in a translucent aerogel that provide U-values from 0.45 to 0.85 W/m²K (0.08 to 0.15 Btu/(hr/ft²/°F) and even super-insulating systems with U-values of 0.30 W/m²K with 20% light transmission (for R-20 U-value of 0.05 Btu/(hr/ft²/°F). These glazing systems reduce heat losses and gains by conduction while still permitting daylight inside the space.

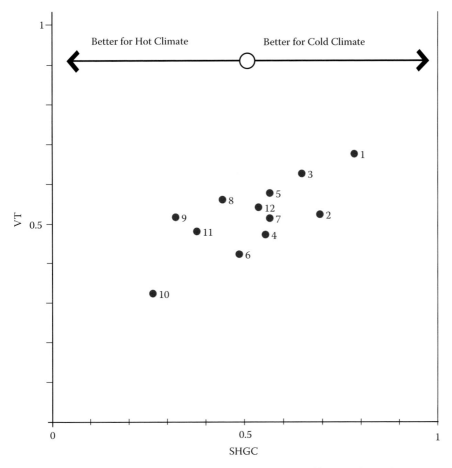

FIGURE 6.36 Visible transmittance and solar heat gain coefficient in different window types.

1. Single-glazed clear aluminum no thermal break
2. Single-glazed bronze aluminum no thermal break
3. Double-glazed clear aluminum thermal break
4. Double-glazed bronze aluminum thermal break
5. Double-glazed clear wood or vinyl
6. Double-glazed bronze wood or vinyl
7. Double-glazed low e wood or vinyl
8. Double-glazed low e wood or vinyl
9. Double-glazed spectrally selective wood or vinyl
10. Double-glazed spectrally selective wood or vinyl
11. Tripled-glazed low e superwindow
12. Triple-glazed clear

6.4.4.8 Orientation of Buildings and Openings

In addition to selecting the appropriate window for each climate and orientation, other strategies can be implemented to reduce the amount of solar radiation that is reaching each transparent component.

The orientation of the windows (and the building) has an important influence on the energy flows, which is also affected by the magnitude of the maximum radiation, the time of day when maximum radiation occurs, the number of hours the surface receives solar radiation, and the ground albedo.

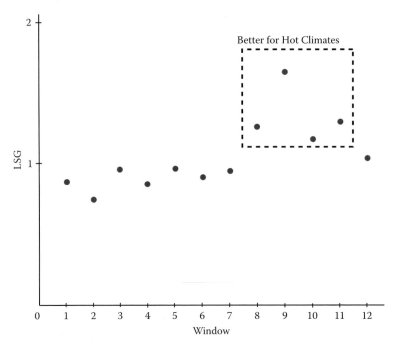

1. Single-glazed clear aluminum no thermal break
2. Single-glazed bronze aluminum no thermal break
3. Double-glazed clear aluminum thermal break
4. Double-glazed bronze aluminum thermal break
5. Double-glazed clear wood or vinyl
6. Double-glazed bronze wood or vinyl
7. Double-glazed low-e wood or vinyl
8. Double-glazed low-e wood or vinyl
9. Double-glazed spectrally selective wood or vinyl
10. Double-glazed spectrally selective wood or vinyl
11. Tripled-glazed low-e superwindow
12. Triple-glazed clear

FIGURE 6.37 Control of solar radiation LSG.

It is usually better to orient buildings with the longest elevations facing toward the north and the south. The equatorial-facing elevation will usually receive more solar radiation in the winter, whereas the polar-facing elevation will receive more solar radiation in the summer, which is not beneficial if cooling is necessary. However, east and west should usually be smaller because even if solar radiation is beneficial in these elevations in the winter it is not helpful in the summer when the sun is more intense during the morning and the afternoon because it is more perpendicular to the façades. These are general rules, though, and the only way to determine the optimum orientation of a project in a specific location is with energy modeling software that uses local temperature and solar radiation data to predict energy use or thermal comfort.

In intertropical latitudes, where seasons are not very differentiated and where the sun reaches high altitudes, choosing between south and north is not significant. It is important to avoid east and west orientations that are lit by the sun's rays in the morning and in the afternoon with incidence near to the normal of the facades and thus are more difficult to protect. In these latitudes it is the roof that becomes a critical building component, because the sun is high in the sky during most of the year with an angle of incidence closer to the normal with the roof.

Skylights in the roofs also affect thermal gains. Horizontal toplights allow abundant solar radiation during the whole year in lower latitudes and during the summer in mid latitudes, increasing cooling requirements. Equatorial-facing clerestories are usually better in mid latitudes because they can be easily shaded to permit solar radiation in the winter but block solar radiation in the summer when the sun is higher up in the sky.

Eolic orientation can be very important in some cases, such as coastal and tropical zones with very predictable wind patterns or frequent and intense winds. However, air movement is very unpredictable and can be easily affected by local obstacles. Solar orientation is usually more predictable and more important than eolic orientation.

6.4.4.9 Glazing-to-Surface Ratio

The dimension of the glazed surfaces affects the amount of total solar energy (diffuse, direct, and global) that can penetrate the building. Larger glazed areas increase the potential for gain and for losses by conduction, affecting the interior air temperature, especially during daytime. More energy inside the building leads to higher temperatures, which in turn leads to a larger indoor temperature swing. The glazing-to-surface ratio is usually calculated by dividing the amount of glazing in the elevations by the surface area (e.g., walls or floor). This ratio can also be calculated as a function of different orientations, for example, as a function of the equatorial-facing glazing-to-floor ratio. The effect of this ratio is discussed in more detail in the section on passive heating.

6.4.4.10 Use of Shading Devices

Because solar radiation can greatly impact internal temperatures and can be blocked easily, shading systems are extremely important to control heat gains by radiation through the windows. The types of shading devices and the process to design them are explained in detail in Chapter 5. Solar protection systems can be fixed or operable and internal or external. Fixed systems are rigidly attached to the building without possibility of regulation. Operable systems allow more flexibility because they can be adjusted manually or automatically, but they require more maintenance.

External shading devices in the form of simple opaque devices between the sun and the openings will reduce the amount of solar radiation on the window and inside the building. Interior shading devices such as blinds and interior curtains are not very efficient, because they block solar radiation after it has already entered the space. The radiation blocked by the blinds or curtains is absorbed by them, heating them and also heating the space by convection and radiation. Half of this reirradiation is toward the inside, and half is toward the outside. Because it is longwave, though, the glass of the window blocks it, and this radiation stays inside. The space between the blind and the window is overheated, and the heat of the blind itself increases the MRT above the air temperature. There are many ways to improve the performance of the window by blocking solar radiation and increasing insulation levels, and some are shown in Figure 6.38.

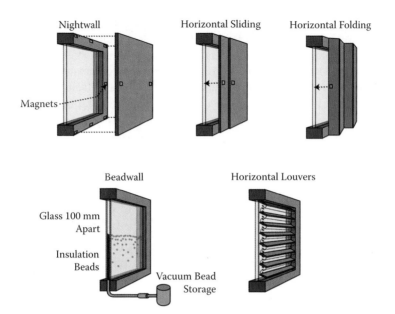

FIGURE 6.38 Different types of movable insulation. (Based on illustrations in Lewis, O. 2001. *A Green Vitruvius: Principles and Practice of Sustainable Architectural Design*, 2nd ed., James and James, London, 2001.)

6.5 HEAT TRANSFER BY CONVECTION

It is important to understand that ventilation can be used for three completely different processes in a building; depending on this process, this exchange of air can be labeled differently. It can be used to supply fresh air, to remove internal heat, and to promote heat dissipation from the body. It can also be deliberate or incidental, which is usually called infiltration. In this section, only the exchange of heat between the interior of the space and the exterior is discussed.

6.5.1 Definition

Convection is the process of heat transfer by flowing and mixing motions in fluids. It is primarily dependent on air temperature. Natural convection occurs when molecules of cool air absorb heat from a warm surface, rise, and carry it away. As a gas or liquid acquires heat by convection, the fluid expands and becomes less dense, rising on top of the denser and cooler fluid. The resulting currents transfer heat by natural convection. The effect of convection on people is reduced by the amount of clothing people are wearing. Layered air spaces in building materials also act as insulation. Heat flow by convection requires a fluid to be present (air), is directional, never convects down (unless it's forced), requires a temperature change, can be blocked completely by obstacles, and is affected by gravity.

Natural convection can be produced in several ways. If there is an air space enclosed by two horizontal elements and the upper element is cold and the lower

Heat Exchange through the Building Envelope

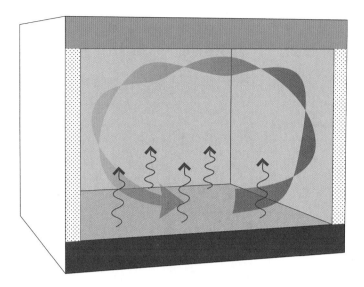

FIGURE 6.39 Natural convection.

element is warm, the air will flow upward and then will descend as it loses its heat in the ceiling (Figure 6.39). However, if there are two horizontal elements and the upper element is warm and the lower element is cold, then the warm air will stay in the upper portion of the space because it is less dense than the cooler air below. Heat will still be transferred from the upper portion to the bottom portion by radiation.

If there are two opposing walls and one is warm and the other wall is cold, the air flows from the warm wall upward and descends down the colder wall, losing its heat. This process can occur in wall cavities when the air gap between the components is large enough (Figure 6.40).

Forced convection occurs when the air is moved by a fan or by the wind and is circulated in a direction in which it would not naturally go.

6.5.2 Air Movement and Infiltration

Heat transfer by convective effects occurs when there is a temperature difference between the internal and external environments of a building (indoor and outdoor). The environment with the highest temperature has more energy and transfers this energy to the area with the lower temperature by ventilation (natural or forced) and infiltration. The amount of heat transferred increases as the temperature difference increases or the air exchange rate increases. The amount of sensible heat transferred is a function of the difference in temperature, the airflow rate, and the volumetric mass of air and can be calculated using

$$Q_v = \rho \times C_p \times v \times \Delta t \qquad (6.27)$$

also expressed as

$$Q_v = 1.2 \times v \times \Delta t \qquad (6.28)$$

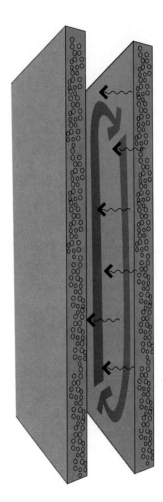

FIGURE 6.40 Convection between vertical elements in a wall.

where

Q_v = amount of sensible heat transferred (W)
ρ = density of air (kg/m³)
C_p = air specific heat (J/kg K).

The product $\rho \times C_p$ = 1.2 kg/m³ × 1,000 kJ/kg K = 1,200 J/kg m³ in locations below 500 m above sea level.

v = air flow rate (m³/s)
Δt = temperature difference between outside and inside $(T_o - T_i)$
T_o = outdoor air temperature (K)
T_i = building's internal air temperature (K)

In the United States with the IP system, this is usually expressed as

$$Q_v = 1.1 \times v \times \Delta t \tag{6.29}$$

where
- Q_v = sensible heat flow rate by infiltration or ventilation (Btu/hr)
- 1.1 = constant derived from the density of air at 0.075 lb/ft³ multiplied by the specific heat of air and by 60 min/hr (min/ft³ hr °F)
- v = outdoor airflow rate in cubic feet per minute (cfm)
- Δt = temperature difference between the outdoor air and indoor air $(T_o - T_i)$ (°F)

It can be also measured in cubic feet per hours, in which case

$$Q_v = 0.018 \times \text{CFH} \times \Delta t \tag{6.30}$$

where
- Q_i = heat flow rate by infiltration
- CFH = cubic feet per hour of air infiltrated in the space; can be determined using different methods
- Δt = temperature difference between the indoor and the outdoors $(T_o - T_i)$

The airflow rate is estimated with

$$v_r = N \times V/3{,}600 \tag{6.31}$$

where
- N = number of air changes per hour (ac/hr)
- V = volume of internal space (m³)

The relationship between the outside and inside temperatures will vary according to the daily temperature cycles and climatic, microclimatic, and seasonal variations. There will be moments in which the outside temperature is higher than the indoor temperature and moments in which the indoor temperature is higher. Outside air is a source of free heating and cooling that must be used whenever possible to reduce the strain on mechanical heating and cooling systems. During the cooling season, if the outdoor air is cooler than the indoor air, ventilation should be promoted. During the heating season, if outdoor air is warmer than indoor air, ventilation should be promoted. This is much less common, and ventilation is used much more to cool the building by eliminating warmer air. Ventilation as a passive cooling strategy is discussed in more detail in Chapter 7.

6.5.3 Controlling the Exchange of Air

Because ventilation can result in either heat gains or heat losses, it is important to control these exchanges according to requirements and needs. In general, if cooling is required, then the building should be

1. Closed when the outdoor temperature is higher than indoor temperature
2. Open when the indoor temperature is higher than the outdoor temperature

These rules can change; for example, increasing the difference between the indoor and outdoor temperatures before ventilation occurs might be recommended. La Roche and Milne (2004) tested a smart ventilation system in several experimental test cells, and after evaluating several control strategies with different relationships between them, such as air change rates, values for comfort low and comfort high, and amounts of thermal mass, it was found that a higher air change rate during the cooling period increases the amount of heat that is flushed out of the building, lowering the temperature of the mass, and lowering the maximum temperature for the next day. The test cells that had the smart ventilation system, and could increase the air change rate when needed, always performed better than the test cells with a fixed infiltration rate. The system performed better with maximum air change rates of 15 air changes/hour rather than 4 air changes/hour, but satisfactory results were achieved with maximum air change rates of only 4 air changes/hour, probably because of the limited amount of mass inside the test cells. These results are discussed further in Chapter 7.

In hot climates, when cooling is required, there will probably still be some moments in which the outdoor air temperature is lower than the indoor air temperature ($T_i > T_o$). During these periods the building will be losing energy to the outside, a process that can be accelerated by increasing the rate of exchange of air between the indoors and the outdoors. When the outdoor temperature is higher than the indoor temperature, there will be heat gain from the outside, and it is better to reduce heat transfer by convection to the minimum to maintain indoor air quality (IAQ). Thus, to avoid elevated indoor air temperatures, it is necessary to seal the building when the external temperature is higher than the internal temperature ($T_e > T_i$) or to open the building when the external temperature is lower than the internal temperature ($T_e < T_i$). In other words, it must be possible to seal and open the building as necessary depending on the indoor–outdoor relationship. This is usually achieved with windows or fans, which could be automatic or manual.

6.5.3.1 Seal the Building When $T_e > T_i$

When ventilation would cause heat transfer to the building's interior (heat gain), the building should be sealed (no exchange of air). However, even when it is sealed, a minimum supply of fresh air must still be available to eliminate indoor pollutants. Air movement can also be generated inside the building without the need to exchange air using portable or ceiling fans. Fans provide huge energy savings compared with mechanical cooling systems in nonextreme climatic situations. It might also be necessary to dehumidify indoor air to reduce the humidity generated by equipment and people. Kitchen and laundry areas can be kept separate from the rest of the building, and humidity can be extracted with forced ventilation. If air temperature and humidity levels increase beyond tolerable levels, then it is necessary to use mechanical cooling.

6.5.3.2 Open the Building When Outdoor Temperature Is Lower than Indoor Temperature ($T_o < T_i$)

This process dissipates hot air from inside the building by exchanging it with cooler outdoor air. There are two basic types of building ventilation that vary depending on the type of building (low mass versus high mass) and the type of climate (hot and humid versus hot and dry).

In hot and dry climates it is important to maximize the air change rate to promote heat exchanges and cooling of the building structure during the night, when the outdoor temperature is lower. Cooling the thermal mass at night will provide a cool internal surface to absorb heat during the day, keeping it cool.

In warm and humid climates, it is important to maximize air speeds inside buildings to cool the occupants and to produce physiological cooling and more thermal comfort. An increase in air velocity promotes evaporative cooling of the skin and reduces thermal resistance of the air film around the human body. Air movement is helpful only when the outside air temperature is lower than the temperature of the skin (about 34°C).

In hot, humid climates, cross-ventilation is the most effective strategy to promote thermal comfort, both by evacuating the heat that can accumulate inside the building and by creating air velocities that can promote physiological cooling of the skin.

Ventilation takes place by naturally occurring pressure differences that generate an exchange of air between the exterior and the interior. These pressure differences can be caused by wind or by the buoyancy effect created by temperature differences.

When wind reaches the windward side of the building, it is blocked and a positive static pressure zone is created while a negative pressure zone is created at the opposite side of the building. If there are windows on both sides, the difference between the high pressure at the windward side (entry area) and the low pressure on the leeward side (suction zone) will generate a flow of air through the building (Figure 6.41).

In many cases, cross-ventilation can reduce the use of mechanical cooling. With air speeds of 0.57 m/s and 0.85 m/s, reductions in the perceived internal temperature between 2.2 and 2.7°C were attained (Cook, 1986). Other research suggests that a 1 m/s air velocity at the body surface creates a perceived cooling effect of 3 K and 1.5 m/s up to 5 K.

The factors that affect the quality of the design as it relates to the air movement inside the space are as follows:

- The building envelope and the location and design of the internal elements and how they affect the internal air distribution
- The shape of the building and ventilation induction to the inside
- The elements surrounding the building (natural and architectural)

6.5.3.2.1 The Effect of the Skin and Internal Elements on Airflow

Different studies carried out on scale models in wind tunnels have evaluated the relationship among the space geometry, location, and shape of the openings on cross-ventilation, which must be considered during the design process to optimize ventilation. These studies have generated several well-known but useful principles:

- There should be an inlet opening or window and an outlet or exit to maximize wind speed.
- There must be a pressure difference between these two for air to flow through the space.

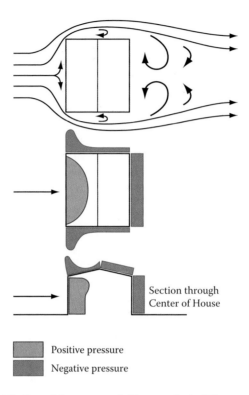

FIGURE 6.41 Distribution of the pressure fields around a building.

- An outlet larger than the inlet allows for more opportunities to direct the flow of air in different directions.
- Different size windows for inlets and outlets affect the air velocity inside the space. Wind speed averages inside the space are higher when both openings are larger. If the inlet is larger than the outlet, the air velocity will increase close to the outlet (Venturi effect). However, this is not usually helpful. If the inlet is smaller, then air velocities will usually be higher close to it, creating more potential for air penetration. A larger difference between the inlet and outlet will generate higher air velocities. Sobin (1981) recommended a ratio of $A_o/A_i = 1.25$ for the highest indoor speed ratios, where A_o = area of outlet and A_i = area of inlet. Higher ratios will produce higher velocities close to the inlet but lower averages across the whole room.
- The location of the outlet does not significantly affect the airflow pattern (Figure 6.42) and the location of the inlet affects the airflow pattern (Figure 6.43). Because the air has mass it has inertia and will continue flowing in the direction that it entered.
- The best location for an inlet is usually close to the vertical and longitudinal center of a wall that is perpendicular to the wind, where the pressure is highest. The best location for the outlet should be in the low pressure zone. The outlet can also be placed high, near the ceiling, to take advantage of a stack effect, if any.

Heat Exchange through the Building Envelope

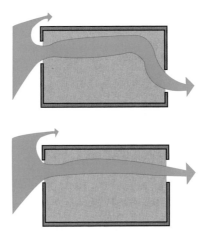

FIGURE 6.42 Location of the outlet affects only the direction of the airflow at the end.

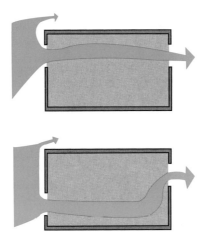

FIGURE 6.43 Location of the inlet affects airflow.

Givoni (1976) suggests that the best ventilation in the space (in terms of overall air movement and homogenous distribution) is obtained when the airflow changes direction inside the room (Figure 6.44) by not aligning the outlet windows with the inlet windows. Sobin (1981), however, found that a change in air flow direction sometimes substantially reduced air velocity.

To achieve sufficient natural ventilation inside the building spaces, it is recommended to have the openings toward the longest facades, which should be facing the wind. Windows should be on both sides of the space with the longer side of the building facing predominant wind directions, creating larger high-pressure and low-pressure zones on either side of the building and shorter routes for the air to traverse (Figure 6.45). Also, it is recommended for the windows to be at least 15% of the floor area to provide the necessary air changes and air movement for comfort, and some codes require all naturally ventilated spaces to be at least 20′ from a window. More

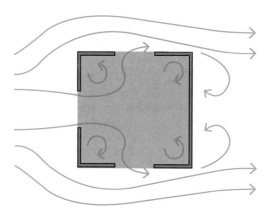

FIGURE 6.44 Wind direction inside a space is affected by the position of the openings.

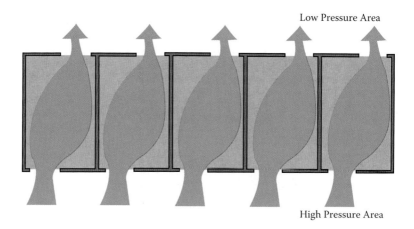

FIGURE 6.45 Aligned rooms can capture more air.

openings in the negative pressure zone of a space usually increase ventilation flow. However, in spaces with higher occupation densities or more internal heat gains, these openings should be larger.

It is always recommended to have windows on at least two external walls of each space. However, when the room has only one external wall, this wall should have two windows, placed at the far ends of the wall to achieve some difference in pressure and internal airflow through the space. It is also possible to increase airflow by changing the pressure around windward windows with wing walls or fins (Figure 6.46). If the wind is hitting facades without openings, most of the other windows will not have sufficient pressure to generate any significant cross-ventilation. In this case pressure differences can be generated with vertical elements such as fins or vegetation that are perpendicular to the wall.

The angle of the inlet to the wind affects the air flow rate. For most of the windows, orientation of inlets at 90° to the wind provides the highest average indoor

Heat Exchange through the Building Envelope

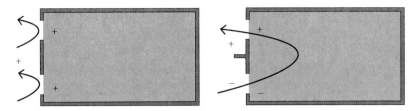

FIGURE 6.46 When windows are on the same façade, fins can be used to increase air velocity.

speed ratios. However, the best ventilation does not always occur when the window is perpendicular to the wind direction. Givoni (1994) suggests that buildings exposed to oblique wind along the windward walls create different pressures in the windows located in these walls, and the air can enter through the upwind window and exit through the downwind window. Certain combinations of inlet characteristics (especially shape) provide good results at 90°, but can also provide equal or better ventilation in oblique (up to 45°) winds than they do in normal (90°) winds. These windows would probably provide better overall ventilation because wind changes direction, and are often probably a better selection than a window that has better performance but with only one direction. Perpendicular winds with cross ventilation in a room produce an average wind speed and internal maximum speed double that of windows on one side only.

6.5.3.2.2 The Effect of the Shape of the Windows on Ventilation

Windows with horizontal proportions generate higher indoor air velocities and catch a larger range of wind directions than vertical shapes (Cook, 1986; Sobin, 1981; Ghiabaklov, 2010). These windows are especially beneficial in locations where changes in wind pattern are more prevalent.

In all cases, windows should be accessible and operable so that users can control and direct winds according to ventilation needs.

6.5.3.2.3 The Effect of Window Obstructions

Internal divisions should not create barriers to inlet openings. If there are partitions, they should be perforated, less than full height, or have openings to permit the flow of air. Fly screens also reduce air speed, especially wire or cotton mesh. Smooth nylon or other plastic screens offer the least resistance. With higher air velocities, a wind screen's effect reducing the air velocity is more noticeable than with lower velocities.

Van Straaten measured the reduction in air velocity produced by screens and found that it depended on the incident wind speed, with a higher percentage reduction when the velocity was lower. For a 0.7 m/s (1.5 mph) wind, the airflow was reduced by 60%, while in a 2.7 m/s (6 mph) wind the reduction was only 28%. This difference is possibly due to the reduction of the wake region behind a cylinder as the Reynolds number increases. Internal airspeeds can only be predicted.

6.5.3.2.4 Building Shape and Ventilation Induction to the Inside

In warm and humid climates, the main purpose of ventilation is to provide maximum penetration and air changes with significant indoor air velocities. In hot and dry climates, the control of ventilation to provide night cooling is the main purpose. The main building elements that affect the ventilation pattern are the height and length of the building and its roof slope, and it is important to consider the following variables in designing for wind:

- An increase in the building's width enlarges the surface exposed to the wind while the depth of the wind shadow or calm zone remains constant.
- An increase in the building's height provides higher speeds in the upper levels but, at the same time, increases speeds toward the first floors. This effect is even more accentuated in buildings with free-flow areas on the ground floor.
- A reduction in the width–height relationship increases the dimension of the wind shadow or calm zone.
- If the roof's slope facing the wind has an angle above 30 degrees, there will be an increase in the depth of the calm zone.
- When the roof's slope is opposite the wind and below 30 degrees, the increase in the dimension of the calm zone will not be significant. An increase in the slope above 30 degrees will not provide any additional calm zone.
- If the slope of a gable roof increases, the height and depth of the wind zone also increases.
- Eaves on the leeward side have little effect on the wind shadow; however, they do affect airflow when facing the wind, contributing to capture the wind. Usually eaves and canopies over the window deflect the indoor airflow upward. If they are separate from the wall, they will increase the air stream and promote air penetration inside the building.
- As the angle between the wind direction and the largest dimension of the building (regardless of the shape) approaches 90 degrees, the calm zone will also increase.

The following considerations should also be taken into account:

- If the openings are at the same level and near the ceiling, most of the airflow will stay toward the top area of the space.
- Openings at the ridges of gable roofs are useful in open-plan buildings because they produce suction effects that allow the hot air to exit.

6.5.3.2.5 *The Effect of the Area Surrounding the Building*

It is important to take advantage of the topography, landscape, and building surroundings to redirect the airflow and to ensure maximum exposure to the breeze. Garden elements such as trees, bushes, fences, or low walls can be used to reconduct breezes and to avoid stagnant air pockets.

If landscaping features are placed very near the building and windows, they may reduce the wind speed and modify the airflow pattern; but if they are placed at the sides of the window, they can act as wing walls, creating a higher positive pressure and increased airflow through the building. Also, wind speeds under tall trees are usually higher than toward the crown.

7 Passive Cooling Systems

All buildings are heated up during the day by effects of solar radiation (sol) and cooled during the night by convection and radiant loss to the sky. In warm climates, the average indoor temperature is generally higher than the outdoor average because the heat gains are usually higher than the losses.

According to Givoni (1994) there is a difference between bioclimatic architecture and passive cooling. Bioclimatic design consists of appropriate architectural design to keep the building from overheating or overcooling and can minimize the difference between the indoor average temperature and the outdoor average temperature but cannot lower the indoor average temperature below the outdoor average values. Chapter 6 discussed strategies for reducing overheating and overcooling in buildings. This chapter deals with passive cooling techniques, which are the only way in a hot climate to achieve average temperatures inside a building lower than outdoors unless mechanical or active techniques are used. In a cold climate we would need active or passive *heating* techniques.

7.1 DEFINITION OF A PASSIVE COOLING SYSTEM

A passive cooling system is capable of transferring heat from a building to various natural heat sinks (Givoni, 1994). To achieve this, the building must have special design details, generally in some part of the envelope. Passive cooling systems provide cooling through the use of passive processes, which often use heat flow paths that do not exist in conventional or bioclimatic buildings.

A building designed using bioclimatic principles is a precondition for the application of passive cooling systems. If the building is not *bioclimatically* designed, then the passive systems will not run efficiently because they will need more energy to cool or heat than if they were already designed efficiently.

7.2 CLASSIFICATION OF PASSIVE COOLING SYSTEMS

Passive cooling systems can be classified according to the different heat sinks that are used (Givoni, 1994): ambient air (sensible or latent), upper atmosphere, water, and undersurface soil. Different cooling strategies can be used to take advantage of these heat sinks (Givoni, 1994):

  Comfort ventilation: providing direct human comfort, by moving ambient air, mainly during the daytime (sensible)
  Nocturnal ventilative cooling: opening the building to cool the structural mass of the building interior by ventilation during the night and closing the building during the daytime (sensible)

Direct evaporative cooling: mechanical or nonmechanical evaporative cooling of air, which is then introduced into the building (latent)
Indirect evaporative cooling: evaporative cooling of the roof so the interior space is cooled without elevation of humidity (latent)
Radiant cooling: transferring heat from the building to the upper atmosphere during the night hours.
Soil cooling: cooling the soil below its natural temperature in a given region and using it as a cooling source for a building by some type of heat transfer mechanism.
Cooling of outdoor spaces: cooling techniques that are applicable to the air of outdoor spaces adjacent to a building, usually using some sort of water evaporation or some sort of expenditure of energy.

This classification is useful because it relates the building systems with the heat sinks (Table 7.1). The applicability of a given cooling system depends on the specified limits of the indoor climate, which varies with the type of building and outdoor climate.

According to Cook (1986), the term *passive* as applied to the integral heating and cooling of buildings was invented in the United States. It generally describes a form of space conditioning without the use of parasitic power sources. Passive systems are driven primarily by natural phenomena. While conventional solar heating and cooling systems for space conditioning are active in their uses of motorized mechanical components to move fluids and air, these buildings are passive because they use few conventional sources of energy.

Passive space heating is driven only by the sun and was named first. *Passive cooling* is its counterpart. While passive heating uses the sun as a single solar heat source, passive cooling embraces several heat sinks and a wide variety of bioclimatic practices in building design, so it is more complex and harder to classify. Passive cooling is implemented when temperatures are above the comfort zone (to the right) and the building gains more heat than it loses.

Not all passive cooling or heating systems are applicable in all types of climates, and it is important to understand the most favorable climate variables. The implementation of each of the systems depends on specific climate conditions such as

TABLE 7.1
Relation of the Heat Sinks with the Passive Cooling Strategies

Building System	Heat Sinks
Comfort ventilation	Ambient air
Nocturnal ventilative cooling	Ambient air
Direct evaporative cooling	Ambient air
Indirect evaporative cooling	Ambient air
Radiant cooling	Deep space
Soil cooling	Earth
Cooling of outdoor spaces	Ambient air or water

Passive Cooling Systems

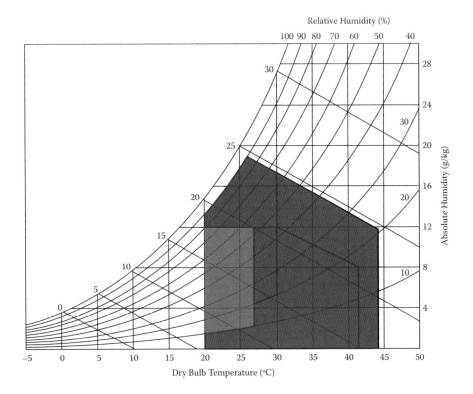

FIGURE 7.1 Outdoor conditions for which passive cooling strategies will help achieve thermal comfort.

relative humidity (RH), temperature, and radiant conditions. Figure 7.1 shows the areas in which passive cooling would be useful. These conditions depend on the nature of the process; some systems will work better under some conditions. It is important to understand when it is appropriate to use each system. The best way to determine when it is more beneficial to implement a particular strategy is to use Givoni's psychrometric chart with the passive strategies overlaid on it (Figure 4.11).

7.3 AMBIENT AIR AS A HEAT SINK (SENSIBLE COMPONENT)

The atmosphere is also a medium of heat transfer, primarily through convection. Cooling by ventilation to the atmosphere is the simplest way to remove heat from buildings, and it is the most common passive cooling method. Indirect and direct evaporative cooling systems are also included in this general group.

Ventilation can be used for three completely different functions in a building: (1) cooling the human body, (2) cooling the structural mass of the building, and (3) indoor air quality (IAQ). Each of these has different requirements that affect the building and the ventilation systems. Good indoor air quality is usually an indirect positive product of ventilation systems. Comfort ventilation is an additional resource that can be implemented to extend the comfort zone to temperatures above the limit for still air. When outdoor air temperatures are lower and more suitable for sensible cooling, structure cooling can be put into effect, usually as nocturnal ventilative cooling.

7.3.1 Comfort Ventilation

Comfort ventilation provides direct human comfort. This is the simplest and most common strategy to improve comfort, and humans have historically relied on it for generations. Comfort ventilation is applicable when still air temperatures seem to be too warm, because air movement extends the comfort zone upward beyond the limit for still air. This air motion can be generated by ventilating the space with outdoor air but can also be produced by circulating air inside the space using a ceiling or portable room fan. The area in the psychrometric chart in which we can achieve thermal comfort by providing natural ventilation is indicated in Figure 7.2.

Using the air as a cooling source may be as simple as opening a window, but "designing" the window for the necessary amount of airflows and air change rates can be a more complicated task. Effective interior convective air motion rates are seldom more than 1/10 of the speed of the natural outdoor motion, even if the window is properly designed (Cook, 1986). It is usually recommended to have air speeds between 10 and 45 ft/min (.06–.25 m/s), regardless of whether the air is replaced or recycled, to achieve a sense of freshness in the space.

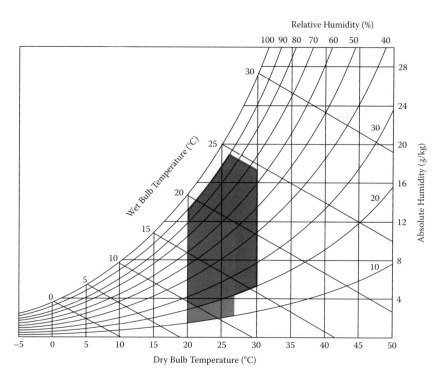

FIGURE 7.2 Applicability of ventilation to achieve thermal comfort.

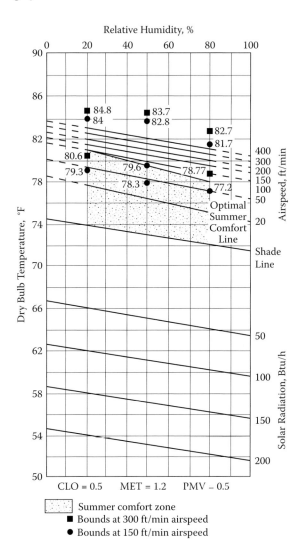

FIGURE 7.3 FSEC comfort chart, showing change in optimal comfort line for various air speeds. CLO, MET, and PMV refer to clothing level, metabolic rate, and predicted mean vote, respectively. (From Cook, J., *Passive Cooling*, MIT Press, Cambridge, MA, 1986.)

Increasing the air motion raises the boundaries of the comfort zone but in a nonlinear fashion. According to the Florida Solar Energy Center, there is a 4°F benefit with air velocities between 100 and 150 fpm (.57 and .85 m/s) compared with still air. Figure 7.3 describes the optimal summer comfort lines for various airspeed levels for rooms where the mean radiant temperature (MRT) equals the dry-bulb temperature (DBT) and where persons clothed in summertime clothing (clothing level [CLO] = 0.5) are performing light office work (metabolic rate [MET] = 1.2). The optimal comfort lines are drawn between the relative humidity lines of 20 and 80%, which are the limits set to reduce health problems and the generation of mold and mildew.

The center of the comfort zone with still air [air velocity below 20 ft/min (0.1 m/s)] and 60% RH will be at 76°F (24°C). The shaded zone indicates the comfort zone where 80% of the population will be comfortable under still air conditions. The 80% comfort zone boundaries for an airspeed between 150 and 300 ft/min (0.8–1.5 m/s) can be obtained by connecting the dotted and square symbols.

The disadvantage of comfort ventilation is that it does not lower indoor air temperature but counts on increasing comfort by elevating sweat evaporation from the skin. Air movement must be constant, especially when air temperatures are outside the comfort zone. The temperature up to which the body can still be comfortable limits the applicability of comfort ventilation. Therefore, ventilation can be applied only when indoor comfort can be experienced at the outdoor temperature with acceptable indoor airspeed. Assuming that an indoor airspeed of 1–1.5 m/s (200–300 ft/min) can be achieved either by natural passive or by active ventilation, Givoni (1998) proposes this outdoor value to be between 27°C and 32°C (80.6°F–89.6°F). Above this temperature air movement will not be useful for comfort, and as humidity goes up it becomes more difficult to achieve thermal comfort with the same air velocity. Thus, the upper limit of the comfort zone goes down as the air temperature goes up (Figure 7.2 and Figure 7.3). I assumed a practical upper limit of around 30°C to account for inefficiencies in the provision of ventilation.

Comfort ventilation is the most common strategy in a warm and humid climate. The archetype of a building for a hot and humid climate is a building that has low mass and large windows to maximize air movement. But even heavy-mass buildings that are continuously ventilated could have lower maximum temperatures than low-mass buildings in many hot climates with different levels of humidity (Givoni, 1998), even though thermal lag could be a problem.

Comfort ventilation can occur by pressure differences acting on inlets and outlets (natural ventilation) or by mechanical means (forced ventilation). In many places, because of its unpredictability, the wind cannot be relied on to provide adequate air motion. In these cases, fans can be used to generate air movement inside the building or to increase the exchange of air between the indoors and the outdoors. Ceiling fans can easily generate air speeds above 200 fpm, and whole-house fans can generate up to 20 air changes per hour. An effective strategy would combine natural ventilation to provide air exchange (building cooling) with ceiling or portable fans for air motion (people cooling) if natural ventilation could not provide sufficient air velocities. The fans can continue to operate, providing air movement, and can reduce the cooling load when air conditioning is needed (Figure 7.4).

7.3.2 Nocturnal Ventilative Cooling

Nocturnal ventilative cooling occurs when an insulated high-mass building is ventilated with cool night outdoor air so that its structural mass is cooled by convection from the inside, bypassing the thermal resistance of the envelope (Figure 7.5). During the daytime, if there is a sufficient amount of cooled mass and it is sufficiently insulated from the outdoors, it will act as a heat sink, absorbing the heat penetrating into and generated inside the building and thus reducing the rate of indoor temperature rise (Figure 7.6). This ventilation system can be either fan forced or

Passive Cooling Systems

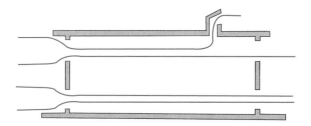

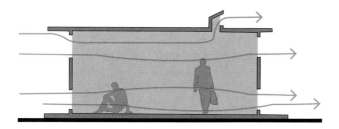

FIGURE 7.4 Windows provide air movement through the space at different heights.

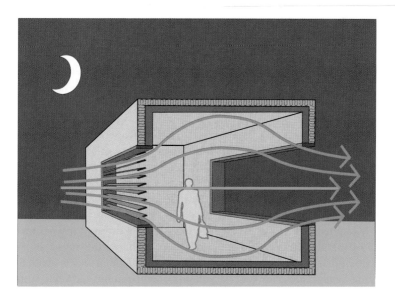

FIGURE 7.5 Nocturnal ventilative cooling (nighttime).

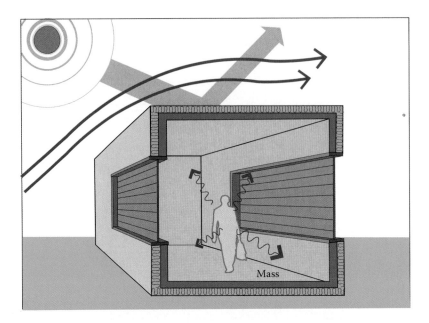

FIGURE 7.6 Nocturnal ventilative cooling (daytime).

natural, through windows that are opened and closed at appropriate times. During overheated periods, the ventilation system (windows or fans) must be closed to avoid heat gains by convection. Nocturnal ventilative cooling is a well-known strategy that has been used for many years, mostly in warm and dry climates. In these climates, buildings with thick walls were cooled during the night and acted as heat sinks during the day. Night ventilation reduces internal maximum temperatures, peak cooling loads, and overall energy consumption and has been well documented (Allard and Santamouris, 1998; Cook, 1989; Givoni, 1994; Santamouris and Asimakopoulos, 1996; Grondzik, Kwok, Stein, and Reynolds, 2010).

The applicability of nocturnal ventilative cooling is limited to a certain range of conditions (Givoni, 1994), which are a function of the needs of the occupants and climatic conditions. Occupants affect decisions such as opening or closing the windows during the night and the desirable comfort levels. The climatic parameters that determine the effectiveness of nocturnal ventilative cooling are the minimum air temperature, which determines the lowest temperature achievable; the daily temperature swing, which determines the potential for lowering the indoor maximum below the outdoor maximum; and the water vapor pressure level or humidity, which determines the upper temperature limit of indoor comfort with still air or with air movement (Geros, Santamouris, Tsangasoulis, Guarracino, 1999). Since the outdoor daily temperature swing usually increases as the air humidity is reduced, the humidity of the air is one of the practical determinants of the applicability of different ventilation strategies. Night ventilation is applicable in regions with a diurnal temperature swings above 15°C (27°F), daytime temperatures between 31 and 38°C (90–97°F), and night temperatures below 20°C (68°F) (Figure 7.7). If the maximum temperature is above this value, there is not enough swing, or it is too warm at night,

Passive Cooling Systems

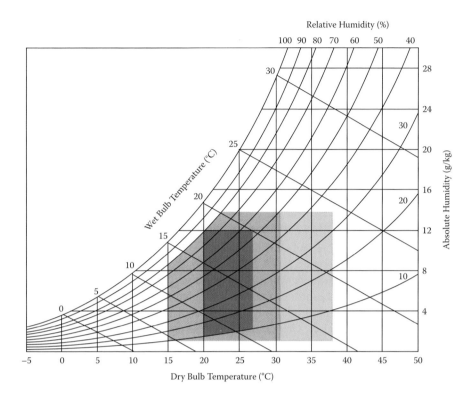

FIGURE 7.7 Applicability of night ventilation with thermal mass (green) or thermal mass only (orange).

then other strategies should be applied during the hottest hours (Givoni, 1994). These conditions are typical of hot and dry climates, making night ventilation one of the preferred strategies in these climates. Even though this strategy is usually not considered effective in warm humid climates, Machado and La Roche (1999) explored the implementation of night ventilation as a passive cooling option for buildings in warm, humid climates.

The psychrometric chart with the area delimited for night ventilation can serve as a general indicator of applicability. The main parameters that determine the performance of night ventilation can be classified in three broad groups: climatic parameters, building parameters, and technical parameters of the technique (Blondeau, Sperandio, and Allard, 1997).

In relation to the climatic parameters, several authors have proposed calculation methods, predictive equations, or rules of thumb to predict performance with different amounts of mass, ventilation rates, temperature swings, or outdoor average temperature. Givoni (1995, 1998) stated that the indoor maximum temperature in night-ventilated buildings follows the outdoor average temperature and proposed formulas to predict the expected indoor maximum temperature with different amounts of mass and insulation. Grondzik et al. (2010) proposed rules of thumb and calculation methods to determine the amount of heat that can be removed from a building with a given amount and distribution of mass, during a typical day, for specific design conditions. Shaviv, Yezioro, and Capuleto (2001) proposes a tool to predict the decrease in the maximum temperature from the diurnal temperature swing as a function of the night ventilation rate and the amount of mass. Computer programs like Home Energy Efficient Design (HEED) allow users to determine the effects of mass, ventilation rates, and insulation on the indoor air temperature or the amount of air conditioning required to maintain a specified comfort temperature (Milne et al., 2005).

In relation to building characteristics, the requirement is usually a minimum amount of thermal inertia, generally defined as a minimum amount of building mass required for thermal storage (Givoni, 2011). More thermal-structural mass increases the efficiency of the technique as the inertia of the building is increased, and the effect of night ventilation can be observed in the next day's lower and delayed indoor temperature profiles (Geros et al., 1999). A useful rule of thumb is that the thermal mass in hot and dry regions with daily temperature swings of 15–20°C can reduce the indoor maximum temperature 6–8°C below the outdoor maximum (Givoni, 1998). The interior planning of the building also plays a very important role, determining the obstruction level of the airflow of the air through the building (Geros et al., 1999).

The technical parameters related to the efficiency of night ventilation deal with the operation period and air change rate. There is agreement on the fact that heat gains by conduction through the building fabric, solar gains through window glazing, infiltration from warm outdoor air, and internal gains from equipment and occupants must be reduced. It is also well known that higher air changes are better than lower air changes, but there is no agreement on the ideal air change rate, and values range from 8 to 20 air changes per hour (Blondeau et al., 1997). If there is sufficient airflow, indoor temperatures close to the outdoor temperature can be achieved.

The operation of the ventilation systems in buildings (windows or fans) is usually controlled by timers or the occupant of the space who must rely on their experience or by thermostats that measure only indoor temperature. Little has been done on techniques to improve the performance of these ventilation controllers. Eftekhari and Marjanovic (2003) proposed a fuzzy controller that monitors outside and inside temperatures together with wind velocity and direction to open or close a window at various degrees to ventilate a building.

Even though their potential for cooling is well known, natural and hybrid ventilation strategies have not been investigated in the United States to the same extent as in Europe (Spindler, Glicksman, and Norford, 2002). This is partly because typical buildings in the United States, especially houses, do not have much mass to act as heat sinks, and the additional mass needed for ventilation is also associated with an increase in costs.

Because heat transfer by convection uses air as the transfer medium, it is proportional to the temperature difference between the interior and exterior of a building and the airflow rate. Equation (6.27) can be used to estimate the amount of heat transferred by convective effects.

If the outdoor temperature is higher than the indoor temperature ($T_o > T_i$), heat will flow to the interior of the building, thus heating the space; and if outdoor temperature is lower than the indoor temperature ($T_o < T_i$), heat will flow to the exterior

Passive Cooling Systems   231

of the building (heat loss to the outside), thus cooling the space. As the temperature difference or the airflow rate between the interior and the exterior increases, the amount of heat transferred between the indoors and the outdoors will increase. Thus, if a building is to be cooled when the outdoor temperature is lower than the internal temperature, it must be designed to have the maximum amount of air changes per hour. The factors that affect the airflow in a building are mostly related to the size and position of the openings.

Equation 6.26 indicates that to prevent the building from overheating in hot climates, it is important to block heat gain from natural or forced ventilation and infiltration. This usually means that in warm climates, the building must be closed when the outdoor temperature is higher than the internal temperature ($to > ti$) and must be opened when the outdoor temperature is lower than the internal temperature ($to < ti$). However, factors such as the need for physiological cooling due to a high relative humidity of the air can also affect the decision to open or close a window.

7.3.3 SMART VENTILATION

La Roche and Milne (2001, 2002, 2003) developed and tested at the University of California, Los Angeles (UCLA) a smart controller that optimizes the use of forced ventilation for structure cooling in a building. This controller used a set of decision rules to control a fan to maximize indoor thermal comfort and minimize cooling energy costs using outdoor air, the greatest potential source of free cooling energy in most California climates. This controller knows when to turn the fan on and off to cool down the building's interior mass so that it can "coast" comfortably through the next day. Thus, the need for air conditioning can be greatly reduced or even eliminated. But the performance of this system can be seriously compromised if the previously mentioned design considerations are not taken into account regarding the amount of mass and the control of solar radiation. The effects of these variables on the system were determined under different air change rates.

The experimental system consisted of a microprocessor controller connected to thermistors that measured temperature, a laptop computer connected to it that contained the control programs and collected and stored experimental data, the two test cells, and an active ventilation system that consisted of a 4-inch inlet and on the outlet side a 4-inch constantly running fan. The inlet damper was opened and almost closed in the experimental cell by a signal from the microcomputer, and the damper was fixed in an "almost" closed position on the control cell to also allow the same controlled amount of infiltration in each cell (0.7 air changes per hour).

Two identical test cells were built simulating the characteristics of typical California slab-on-grade houses. Only the characteristics that would affect the thermal performance of the ventilation system were included in the cells: the insulation level, the brick slab, and the glazing. Another simplification was that the cells had only a south-facing window so that they would receive the same amount of radiation at the same time. The cells were 122 cm (4 ft) wide by 244 cm (8 ft) long and 244 cm high and were oriented with the longest facade toward the east and west. The cells had 7.6 cm (3 in.) foam R12 insulation on the outside and 6.4 mm (¼ in.) gypsum board inside the walls and roofs with a U-value of 0.43 W/m²K in the walls. The east and west walls had additional shading provided by an insulation panel separated 8 cm from the wall. A calibration series, with both cells under identical conditions, demonstrated that these panels eliminated the distortion caused by solar radiation in the morning and afternoon. The floor was hardboard placed on top of an insulation panel, and the roof had two layers of insulation with a U-value of 0.22 W/m²K. There was a 61 × 61 cm (2 × 2 ft) double-pane window on the south side with a solar heat gain coefficient (SHGC) of 0.72 and a U-value of 4.25 W/m²K (0.75 Btu/h ft² °F). The area of the window was 0.37 m² (4 ft²) for a ratio of the glazing-to-floor area of 12.5%. The walls and windows were carefully sealed so that infiltration was controlled only by the fan and damper system.

One of these cells, with the smart operable venting system, was the "experimental" cell in which the air change rate varied between 0.7 and 3.9 air changes per hour. The other cell was the "control," with a fixed ventilation rate of 0.7 air changes per hour (Figure 7.8).

Different control rules were tested to determine which was more efficient using outdoor air to achieve thermal comfort. The controller rules that were tested were simple enough so that they could be built into a thermostat. Four variables were used to calculate the appropriate air change rate: indoor air temperature, outdoor air temperature, comfort low, and comfort high. The first two measure the air temperature and the need for cooling. If the indoor temperature was higher than the outdoor temperature, then the airflow rate increased by opening the damper; and if the indoor temperature is higher than the outside temperature, the damper was closed to reduce the airflow rate. Comfort low and comfort high set the upper and

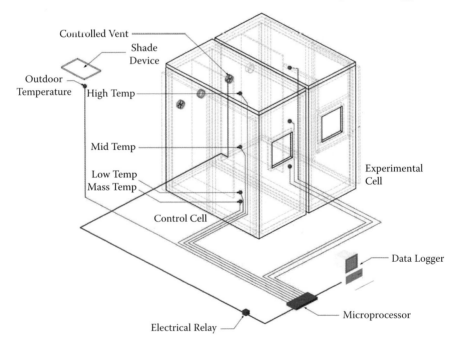

FIGURE 7.8 Experimental setup for the smart ventilation system.

Passive Cooling Systems

lower limits of the comfort zone and determined the air change rate accordingly. If the air temperature was above the comfort zone, it was too warm to provide cooling; and if it was below the comfort zone, no additional cooling was needed. A wider comfort dead band reduced the energy needed for cooling.

Various control strategies were tested with different relationships among the variables (some rules did not use all variables), air change rates, and values for comfort low and comfort high, in summer 2000 and 2001, and are discussed in more detail in La Roche and Milne (2001, 2002, 2004). The rule that achieved the most hours in comfort and the lowest maximum temperatures in the experimental cell is expressed as

$$\text{If } T_o < T_i \text{ and } T_i > C_{f\_low} \text{ and } T_i < C_{f\_high} \text{ then fan ON else fan OFF} \quad (7.1)$$

where

T_o = temperature outside
T_i = temperature inside
C_{f_low} = comfort low at 18.33°C (65°F)
C_{f_high} = comfort high at 25.55°C (78°F)

This is the rule that was then used to determine the effects of the mass and window size. A snapshot of one of the series is shown in Figure 7.9.

Research with these test cells at UCLA demonstrated that it was possible to use smart controllers to cool a test cell in Los Angeles and thus to have a building with reduced internal loads. The factors that affect the performance of this system are the air change rate, the value of comfort low, the thermal capacity of the building, and solar gains through the windows.

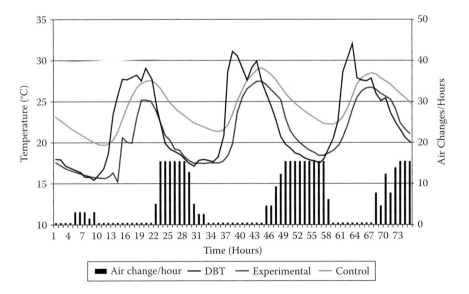

FIGURE 7.9 Temperature and air changes in an experimental series in the smart ventilation system.

A higher air change rate during the cooling period increases the amount of heat that is flushed out of the building, lowering the temperature of the mass and the maximum temperature for the next day. The test cells that had the smart ventilation system and could increase the air change rate when needed always performed better than the test cells with a fixed infiltration rate. The system performed better with maximum air change rates of 15 air changes per hour instead of 4 air changes per hour, but satisfactory results were still achieved with maximum air change rates of only 4 air changes per hour.

With a smart controller, not only the lowest outdoor temperature but also the value of comfort reveals the potential for cooling. After the indoor air temperature is cooled to this value, the damper is closed and the indoor air temperature is maintained, even though the outdoor temperature could descend below this point. If the value of comfort low is reduced, the indoor maximum temperature of the building the next day is also reduced. To achieve maximum cooling the next day, the value of comfort low should be set to the minimum temperature that can be tolerated at night; 18.33°C proved to be a useful number, but series with a comfort low of 15°C performed even better.

The amount of mass inside the building affects the thermal inertia so that the cooling effect, which is mostly at night, can be translated to the next day, thereby reducing the maximum temperatures inside the building. Two amounts of mass were tested using the smart controller. With a slab-on-grade building as defined by the California energy code, the difference between the control and the experimental cell is 2.2°C. When more mass is added to the experimental test cell, the temperature difference increases to 3.1°C. Even with additional mass, the rate of 3.9 air changes per hour seemed sufficient to cool the amount of mass used in the experiment to the comfort low value of 18.33°C.

Solar radiation is an important factor that reduces the performance of the cooling system if its incidence inside the space is not controlled. In all the series and with all indicators, an increase in the shaded area of the window size was followed by an improvement in the performance of the system. The performance was inversely proportional to window size, and performance was also consistently better when the smart ventilation system was used instead of the fixed infiltration system.

In all cases, comfort was improved by more mass, smart ventilation controllers, or smaller windows. Even though the systems with "no windows" exhibit the best performance, windows are important sources of natural light and views in buildings, so they obviously must not be completely eliminated. South-facing windows are also important assets for winter heating in mild, mid-latitude climates. Shading systems should be designed to block solar radiation from these windows in the summer and to increase the cooling performance of the system.

This smart controller has the additional advantage of being able to adjust the ventilation rate as needed in the building and when cool enough temperatures are available outdoors. The hours in which the air change rate must be reduced or increased vary from day to day, and the system sometimes turns on during the day if conditions are favorable. A smart controller provides additional cooling compared with traditional systems that have fixed operating times, because it knows when it will be effective to cool with outdoor air. This improves the performance of the cooling system, reducing the amount of mass needed. Furthermore, the cooling system helps to maintain indoor temperatures inside the comfort band, reducing the number of overheated hours compared with the control cell. The system keeps the indoor air temperature inside the comfort zone by using outside air to cool whenever it is possible and using the thermal inertia of the mass to coast during the warmer hours of the day.

Because comfort low and comfort high in the system are also adjustable by the user, the user has control over his environment, and the system can be tailored to individual requirements, thus increasing the controller's performance even more. It would be a simple matter to add a thermistor to read outdoor temperature and to modify the microprocessor of a thermostat to implement these rules in a smart thermostat. It is hoped that these resources will inspire manufacturers to expand their product lines to offer these capabilities.

7.3.4 Effect of Shading on Smart Ventilation

Many buildings tend to be designed with an abundance of windows, which increases losses in the winter but also heat gains by solar radiation in the summer, creating uncomfortably warm indoor conditions when mechanical cooling is not used or increasing the energy used by mechanical cooling systems when it is used. Blocking this solar radiation in the summer reduces energy consumption and indoor temperatures. The same microprocessor controller that controls the fan is also tested to control automatic window shades and to minimize cooling loads. Commercially available electrically operated venetian blinds were used in this study. Already on the market, especially in Europe, are extendable and retractable awnings, vertical external operable louvers, internal operable draperies, and venetian blinds. However, these are not usually used with a microprocessor thermostat that can read indoor and outdoor temperatures and operate shading devices to optimize indoor temperatures.

As before, data from both the experimental cell and the unshaded control cell are automatically recorded every few minutes. These data can be input into spreadsheets and the results plotted as the experiments are in progress. The controller measures indoor temperature/outdoor temperature, and in these series, surface temperature of a metal plate in the south wall, which was used as a rudimentary detector of solar radiation. It also is preset with the upper and lower limits of comfort temperature. The controller also needs to know whether it is the heating or cooling season; however, our simulation studies imply that this can be calculated on the basis of average temperatures during the prior few days (Figure 7.10).

Summer Rule: In the summer, two conditions must be satisfied to close the shade. The first condition is that shading is provided whenever the indoor temperature is above 70°F, which is assumed as the shade line. The second condition is that a black south-facing metal plate (the window is also facing south) must be warmer than the air temperature. This indicates that the sun is facing that window and is receiving solar radiation.

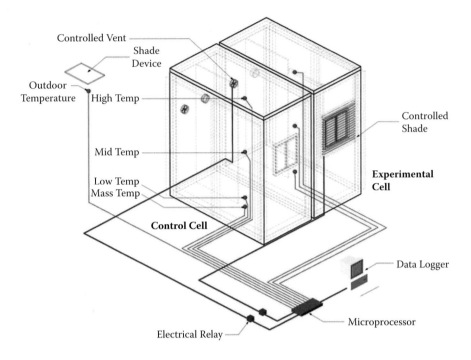

FIGURE 7.10 Experimental setup for the smart shading and ventilation system.

Winter Rule: The shade system is used to avoid overheating of the building through the windows. The shade system is set so that the louvers will close if the indoor temperature is higher than a specified value, in this case 75°F (23.9°C), or 3°F lower than the value set for comfort high (78°F).

Many series were performed to optimize the performance of the different systems, and only one of these is presented as an example (Figure 7.11). Several equations were also developed to predict the performance of these systems (La Roche and Milne, 2000, 2001, 2002, 2003, 2004a, 2005a).

7.3.5 Alternative Methods to Night Ventilate: Green Cooling

A living or green roof is a roof that is substantially covered with vegetation. Green roofs have been proven to have positive effects on buildings by reducing the stress on the roof surface, improving thermal comfort and reducing noise transmission into the building, reducing the urban heat island effect by reducing "hot" surfaces facing the sky, reducing storm water runoff, reoxygenating the air and removing airborne toxins, and recycling nutrients and providing habitat for living organisms, all of this while creating peaceful environments.

The positive thermal effects of green roofs are usually described by the reduction of the external surface temperature due to the effect of vegetation and the reduction of the thermal transmittance of the assembly, mostly due to the effects of insulation, usually placed between the sustaining material and the interior space of the building.

Passive Cooling Systems

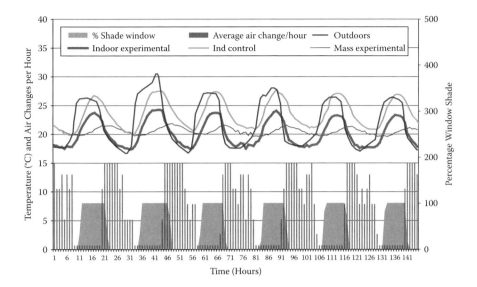

FIGURE 7.11 Example series with smart ventilation and smart shading.

A green roof without this added insulation has a low thermal resistance but does have thermal mass. Apart from providing protection against overheating, a green roof can also provide some cooling through the evaporative process in the plants (Del Barrio, 1998). Their vegetative matter absorbs solar radiation through the biological processes of photosynthesis, respiration, transpiration, and evaporation. However, the solar radiation that bypasses these processes can seep into the building envelope (Niachou, Papakoustantinou, Santamouris, Tsangrassoulis, and Mihalakakou, 2001).

Studies have demonstrated that a well-planned and managed green roof—with insulation—acts as a high-quality insulation device in the summer (Theodosiou, 2003). But little has been done to take advantage of the mass of the green roof as a heat sink in temperate or hot climates. By reducing daily thermal fluctuations on the outer surface of the roof and increasing thermal capacity in contact with the indoors, green roofs can contribute to the cooling of spaces if the mass of the soil is cooled. Some research in this direction indicates the potential for reducing the cooling loads inside buildings (Eumorfopoulou and Aravantinos, 1998; La Roche, 2006c).

During the course of three summers, several green roofs were built and tested at the Lyle Center for Regenerative Studies at Cal Poly Pomona. The university is located about 30 miles east of Los Angeles in southern California, in a hot and dry climate with mild winters (Figure 7.12).

This section focuses on the cooling potential of green roofs combined with night ventilation by looking at the internal temperature of experimental test cells and in a full-size residential building.

Three test cells with a dimension of $1.2 \times 1.2 \times 1.2$ m were built using 2×4 in. stud wall construction with drywall on the inside, plywood on the outside, and batt insulation in between for a U-value of 0.12 W/m² K. The exterior is white, and the three cells have 0.61 m × 0.61 m (2 ft × 2 ft) single-glazed windows facing south that

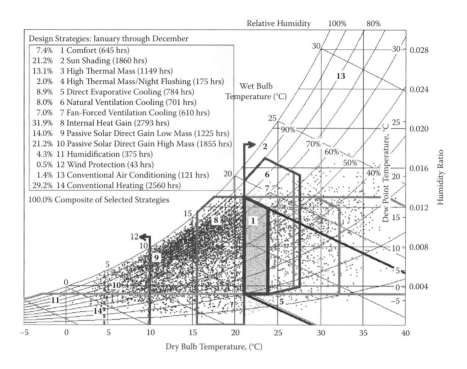

FIGURE 7.12 Pomona climate plotted in Climate Consultant.

were replaced by double-glazed windows in 2007 in the last series. All of the cells have 3.8 cm thick concrete pavers as the slab (Figures 7.13 and 7.14).

The roof is the only difference between the cells. The first cell has a code-compliant insulated roof, with a U-value of 0.055 W/m² K, painted white; the second cell has an insulated green roof; and the third one has an uninsulated green roof (Figure 7.14). The growth medium in the uninsulated green roof is thermally coupled with the interior via a metal plate, while in the other green roof there are 10 cm of matt insulation between the space and the soil. Night ventilation is provided with a

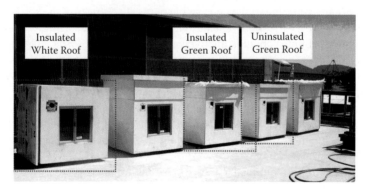

FIGURE 7.13 Photo of experimental test cells at the Lyle Center for Regenerative Studies at Cal Poly Pomona.

Passive Cooling Systems

FIGURE 7.14 Section through experimental test cells used in the green roof experiments.

fan, and all of the cells are equipped with dimmers and timers to adjust the ventilation rate and start–end times.

The green roofs were of the extensive type. Two types were built. The first one, in 2005, was covered with Saint Augustine grass above a layer of soil 7.5 cm thick with 2.5 cm of gravel and a plastic liner underneath. Drainage tubes were spread through the gravel with perforations that capture the excess water and drain it outside the building. The plastic liner is spread above a metal plate, supported by wooden joists. The metal plate assures thermal coupling between the mass and the space underneath (Figure 7.14). The green roof in the cells was substituted by another one in 2007, designed so that it could be built on a large scale with minimal technology, low-cost materials, and little maintenance (Hansanuwat, Lyles, West, and La Roche, 2007). "Rice sack" tubular bags developed by the Cal Earth Institute were cut and filled with a growth medium containing 50% native soil and 50% perlite. These bags were placed above an impermeable layer of plastic, which was placed on top of the roof decking as a moisture barrier. Slits were cut into the topside of the bag, and sedums and succulents were planted, which have little need for water, maintenance, or soil depth. The bags were photodegradable to sunlight; thus, over time the surface of the bags disappeared and created a soil strata evenly distributed over the roof. This same system was also used in an affordable low-cost house prototype for Tijuana, Mexico, that was tested in fall 2008.

Many series were performed, one of which is shown as an example in Figure 7.15. This series initiated on September 7, 2005; there is no window. During the day, the values of the maximum temperatures in the insulated control cell are an average of 0.7°C below the outdoor temperature; in the insulated green roof they are 2.6°C below the outdoor temperature; and in the uninsulated green roof they are 3.6°C below the outdoor maximum temperature (Figure 7.15).

In 2008 the single-glazed window was substituted by a double-glazed window, shaded with a 60 cm overhang that provided complete shading during the summer noon hours and partial shading during the morning and afternoon hours. The Saint Augustine grass was substituted in 2007 by a succulent species that needed little water. Results of tests in 2008 are consistent with previous series and also indicate that the green roof with no insulation performs better than the insulated green roof and the insulated white roof (Figure 7.16).

An uninsulated green roof combined with night ventilation can help cool a space in two ways: (1) the canopy layer reduces the effect of solar gains by reducing the sol–air temperature, and (2) the growth medium acts as a heat sink.

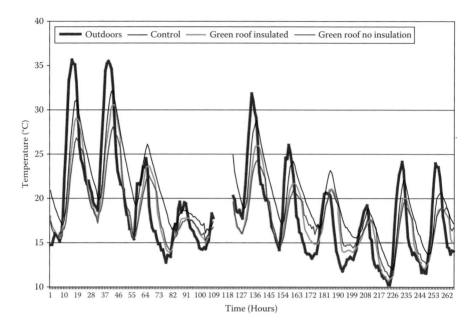

FIGURE 7.15 Control, insulated, and uninsulated green roofs series in 2005.

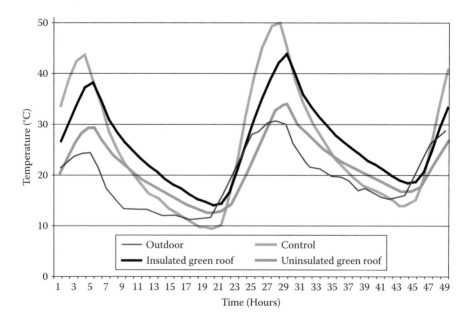

FIGURE 7.16 Two days in the 2008 green roof series.

Results of tests at the Lyle Center for Regenerative Studies, over several years with different types of plants, windows, and shading systems, consistently indicate that during the summer in this climate the uninsulated green roof performs better than the insulated green roof and the insulated white roof. Thus, in a warm or mild climate with cool nights, it is possible to combine a green roof with night ventilation, coupling the soil layer with the interior of the building. The vegetation in the canopy layer improves the performance of the system by blocking solar gains.

Simple equations are derived from the experimental work to calculate internal maximum temperatures as a function of outdoor maximum temperature, daily swing, and glazing-to-floor ratio. One of these equations is tested in a real-size building with acceptable results on cloudy days. More series should be performed under different climates and with different types of buildings to determine the applicability of insulated and uninsulated green roofs and the effect of volume, thermal lag, thermal capacity, and other building variables. Until these series are performed, an applicability range similar to that indicated for thermal mass and night ventilation in Givoni and Milne's chart is proposed (Figure 7.17). In climates with moderate heating needs, where heat loss by conduction is not a critical issue, the uninsulated green roof could probably also be combined with a passive solar heating system (direct or indirect). The lower temperature limit of the climate zone that can be cooled by a night ventilated green roof is a line slightly above

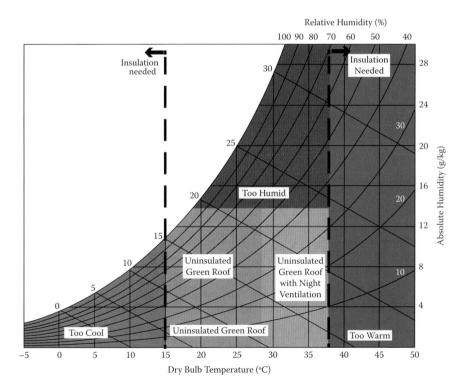

FIGURE 7.17 Applicability of night-ventilated uninsulated green roofs.

the balance point temperature of a building, and the upper limit is determined by the upper limit of the night vented zone with thermal mass. More soil provides more thermal mass, which would affect the upper and lower limits. Caution must be taken in implementing the uninsulated green roofs because even if they do provide benefits when combined with night ventilation in some climates as indicated in Figure 7.17, the losses or gains during other periods can be larger than the benefits provided by the night ventilation cooling. However, in climates in which most of the points are inside the applicability area, green roofs, when combined with night ventilation, can lead to more comfortable conditions inside buildings and to increased energy efficiency. This should give them added value, thus increasing their applicability.

7.4 AMBIENT AIR AS A HEAT SINK (LATENT COMPONENT: EVAPORATIVE COOLING)

In the broadest sense, the term *evaporative cooling* applies to all processes in which the sensible heat in an air stream is exchanged for the latent heat of water droplets or wetted surfaces. Evaporative cooling uses the local atmosphere as a heat rejection resource. The amount of heat absorbed in the process of water evaporation (its latent heat) is very high compared with the other modes of heat transfer, which are common in buildings. This process is adiabatic, which means that no energy is gained or lost. The sensible temperature is reduced with a gain in humidity. When moisture is added to the air, its absolute humidity increases while dry-bulb air temperature decreases. The evaporation of water is the basis of passive cooling systems such as the mechanical evaporative cooler, and on a psychrometric chart this pattern follows the wet-bulb line upward to the left. There are several systems to produce evaporative cooling, and they usually involve a tower to produce the necessary height to evaporate the water and produce cool air or a pond. In hot and dry climates, evaporative cooling also increases thermal comfort because it brings the dry-bulb temperature closer to the comfort zone. In warm and humid climates, the increase in relative humidity is not beneficial because conditions are then usually even more uncomfortable.

The heat loss potential of evaporation of the water is affected by the atmospheric pressure and the quantity of water vapor already in the air. At a standard barometric condition of 29.92 inches of mercury (760 mm Hg), water boils at 212°F (100°C) with a latent heat of vaporization of 1,044 Btu/lb (2.255 kJ/kg). Thus, if evaporation of 1 pound of water requires about 970 Btu (1050 Btu at 76°F) and 12 lbs of water evaporated in 1 hour is equivalent to 1 ton of refrigeration or 12,000 Btu/h, then 1 pound of water will cover a 1 square foot pool to a depth of approximately 3/16 inches. In hot, dry climates, rates of evaporation of about 1 inch a day can be predicted; thus, a 1 ft² open pond can evaporate 5.2 lbs of water per day, which produces cooling of about 5,460 Btu/day or 227.5 Btu/h ft².

An evaporative air cooling system is direct when the air stream comes in contact with liquid water or indirect when the air is cooled without addition of moisture by passing through a heat exchanger, which uses a secondary stream of air or water that has been evaporatively cooled.

7.4.1 Direct Evaporative Cooling

In a direct evaporative cooling system, the air is cooled by evaporation of water, and then the humidified and cooled air is introduced into the building to cool the space. When water evaporates within a stream of ambient air without supply of external heat, the temperature of the air is lowered and its moisture content is elevated while its wet-bulb temperature (WBT) remains constant (refer to Figure 3.8), providing adiabatic cooling. If the temperature as it lowers reaches the saturation curve, condensation will start at the dew point temperature (Figure 7.18, Point B). As the air loses water (Figure 7.18, Points B to C) the air is dehumidified by cooling in an amount that can be determined in the absolute humidity scale.

According to Givoni (1994), direct evaporative cooling can be applied only when the WBT of the ambient air does not rise above about 22°C (72°F), and indirect evaporative cooling can be applied to a WBT of about 24°C. The applicability of direct and indirect evaporative cooling is indicated in Figure 7.19. Direct evaporative cooling is not applicable with air with elevated water vapor content such as in hot and humid regions. Direct evaporative cooling systems do not need to be complicated; the use of wetted pads close to windows or water surfaces like fountains are good examples of direct evaporative coolers.

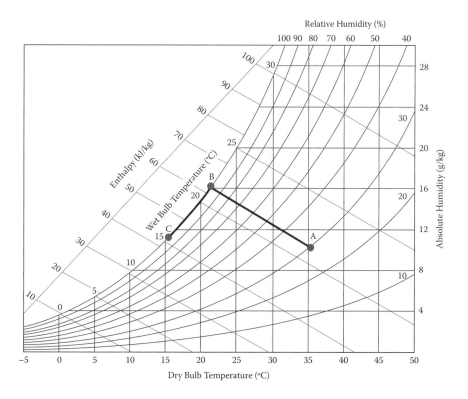

FIGURE 7.18 Adiabatic cooling and condensation.

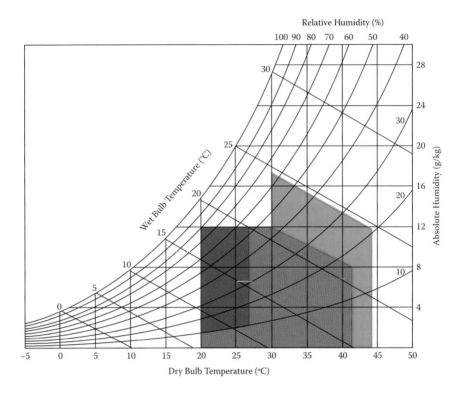

FIGURE 7.19 Applicability of direct and indirect evaporative cooling.

Even though evaporative cooling towers can be used for both direct and indirect evaporative cooling, they are discussed in the section on direct evaporative cooling, and roof ponds are discussed in the section on indirect evaporative cooling.

Cool towers usually have wetted pads at the top or "showers." Cunningham and Thompson (1986) developed a downdraft tower with cellulose pads in the four faces at the top to which water is sent to maintain humidity (Figure 7.20). The hot dry air that enters the top of the tower and passes through the pads, cools, becomes denser, and drops to the base of the tower. The base of the tower is connected to the inside of the space and thus cools it. Excess water is collected at the bottom and recirculated with a pump. Cunningham and Thompson's system had very good performance and achieved exit temperatures of 23.9°C with outdoor temperatures of 40.6°C and WBT of 21.6°C demonstrating the power of cool towers in hot and dry climates. Givoni (1994) took two days of experimental data from their work and developed a formula to calculate the temperature of the air exiting the tower:

$$T_{exit} = DBT - 0.87 \times (DBT - WBT) \qquad (7.2)$$

where
  T_{exit} = temperature of the air exiting the tower in degrees Celsius
  DBT = dry bulb temperature of the air outside (°C)
  WBT = wet bulb temperature of the air outside (°C)

Passive Cooling Systems

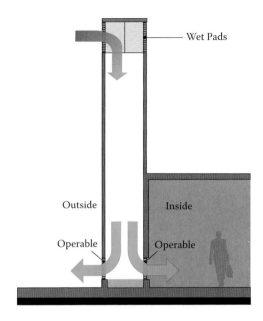

FIGURE 7.20 Cool shower with wetted pad.

This equation does not consider the wind speed. Because there is close agreement between the measured and computed temperatures, it seems that wind speed has only a minor effect on the performance of the tower, making this formula especially useful when wind speed data for the local site is not available.

Givoni (1994) proposes another equation to determine the flow rate and the exit air speed as a function of the area of the cooling pads, the tower height and the wet bulb temperature depression:

$$\text{Flow} = 0.033 \times A_{evap} \times \text{sqrt}(H \times (DBT - WBT)) \quad (7.3)$$

$$\text{Speed} = \text{flow}/A_{tower} \quad (7.4)$$

Flow = flow rate of air through the tower (m^3/s)
Speed = exit air speed (m/s)
A_{evap} = area of the wetted pads) (m^2)
A_{tower} = cross area of the tower (m^2)
H = height of the tower (m)
0.033 = constant representing the total pressure drop through the whole system that includes the cooling tower and the building

This constant changes if the configuration changes. This can happen directly by conduction through the walls or indirectly through the use of heat exchangers that cool the air that penetrates to the building.

Givoni originally developed a cooling tower based on a shower instead of wetted pads for the '92 World Expo in Seville, Spain. In this type of tower there are nozzles toward the top of the tower that spray very fine drops of water downward. As these drops fall, they bring with them a large volume of air, which is cooled by evaporation

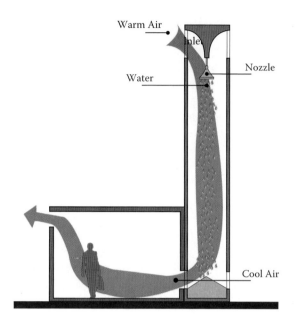

FIGURE 7.21 Direct evaporative cooling with shower.

of the water drops, as well as the water to a level close to the air's WBT. The water is collected at the bottom of the tower and then pumped upward. A wind scoop can be placed above the showerhead to increase the airflow as it goes down (Figure 7.21). Theoretically almost any type of water can be used in these systems; however, there have been some problems with nozzle clogging with hard water.

7.4.2 Indirect Evaporative Cooling

Evaporative cooling can also be passive and indirect, when the system cools a given element of the building, such as the roof or a wall and this cooled element, serving as the heat sink of the building absorbing through its interior surface, the heat inside the building. This heat is transferred in the water to a pond or shower where the water is cooled by evaporation. Thus, the heat from the interior of the building is transferred to the exterior of the space. The advantage of this system is that the interior space is cooled without elevation of the humidity, thus maintaining thermal comfort in more humid climates. Figure 7.22 shows a diagram of an indirect evaporative cooling system with a shower, and Figure 7.23 shows an indirect evaporative cooling system built at UCLA by Sukanya Nutalaya. There, water from the building is transferred in a closed loop and is transferred to the water in the shower via a heat exchanger (Figure 7.22). Because the indoor temperature is not elevated by indirect evaporative cooling and indoor air speed can be augmented by internal fans, the limit for indirect evaporative cooling is about 2°C higher than the upper limit for direct evaporative cooling (Figure 7.19).

Examples of indirect evaporative cooling systems include installing shaded water ponds above uninsulated roofs or simply sprinkling water on the external surface of a roof. In the case of ponds, the surface temperature of the roof follows closely the

Passive Cooling Systems

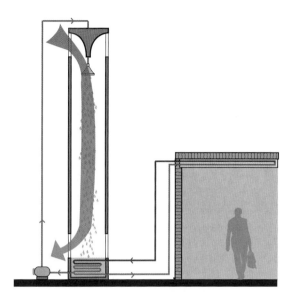

FIGURE 7.22 Indirect evaporative cooling with shower.

FIGURE 7.23 UCLA indirect evaporative cooling with shower.

ambient wet-bulb temperature, while the ceiling, which is thermally coupled to the pond, acts as a radiant or convective cooling panel for the space under it. In a roof pond system the roof can be cooled by radiation, water evaporation on its surface, or both, and the indoor space is cooled by conduction of heat from the inside to the cooler outside. During the daytime, the roof pond is covered and shaded to block solar radiation and to reduce heat gains from the outdoors while it absorbs energy from the inside of the building (Figure 7.24). Water is an efficient heat sink

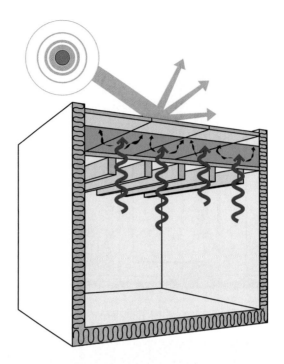

FIGURE 7.24 Roof pond daytime performance.

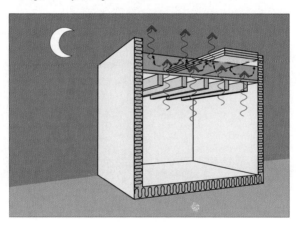

FIGURE 7.25 Roof pond nighttime performance.

that can absorb 1,157 Wh of heat per m³ of volume and degree rise in temperature. If the water is enclosed in containers, it simply works as a heat sink for the space underneath. If the water is exposed, some cooling by evaporation can occur with loss of water. This evaporative cooling effect can be augmented by spraying of water by nozzles, which can be located under the shaded area. During the night, the insulation above the water is open so that the water containers cool by radiation to the night sky (Figure 7.25). A clear sky is needed to maximize the radiant loss. Evaporative

TABLE 7.2
Types of Roof Ponds

	External Protective Cover	
Water	**Fixed Cover**	**Operable Cover**
Water enclosed in containers	The cover is fixed at an angle that blocks solar radiation during the daytime and allows heat loss during the night. Example: La Roche Marnich Cal Poly Pomona (Marnich, Yamnitz, La Roche, and Carbonnier, 2010)	The roof is operable, closed at night and open during the day. The containers are heat sinks during the daytime and radiate the energy to the sky at night. The roof is open at night and closed during the day. Example: Skytherm System
Water enclosed in containers that are sprayed or flooded	The cover is fixed at an angle that blocks solar radiation during the daytime and allows heat loss during the night. Additional cooling can be provided by water sprays. Example: Camelback School, AZ (Yannas, Evyatar, and Molina, 2006)	The roof is operable, closed at night and open during the day. The water is enclosed in containers that cool by radiation at night and act as heat sinks during the day. Additional cooling can be provided by water sprays.
Open water	The cover is fixed at an angle that blocks solar radiation during the daytime and allows heat loss during the night. Nozzles underneath the cover provide additional cooling by evaporation. The cover can also be fixed and cover the water as in a floating insulation system, in which case the water is cooled by radiation and evaporation during the night. Examples: Givoni–La Roche UCLA Univeristy of California Los Angeles (La Roche and Givoni, 2000) La Roche Marnich Cal Poly Pomona (Marnich, Yamnitz, La Roche, and Carbonnier, 2010) Gonzalez Universidad de Zulia (Givoni 2011)	The roof is operable, closed at night and open during the day. The water cools by radiation at night and a heat sink during the day. Additional cooling can be provided by water sprays. Example: Operable louver system at Cal Poly Pomona (Marnich, Yamnitz, La Roche, and Carbonnier, 2010)

cooling with the nozzles can also be implemented at night because even though the wet-bulb depression will be lower, there are no heat gains from the sun. A key feature of a roof pond is that there is no insulation between the water and the space; the insulation or the shading system is placed above the water. Many variations of roof ponds control evaporation and radiation in different forms (Table 7.2). Roof ponds can have fixed or operable shading or insulating covers, and the water can be enclosed in bags or containers. These bags can also be sprayed or flooded for additional evaporative cooling.

7.4.2.1 Givoni–La Roche Roof Pond at UCLA

La Roche and Givoni (2000) tested a roof pond system at the Energy Laboratory of the Department of Architecture and Urban Design at UCLA during summer 1999 (Figure 7.26). This roof pond works as an indirect evaporative system and consists of an insulated water tank, cooled at night, and a test cell with cooled water circulating in pipes embedded in its concrete roof.

The test cell is 1.25 m wide, 1.40 m high, and 2.55 m long. Its walls are built with plywood outside and fiberglass inside. The roof has 100 mm of concrete (tubing embedded) above a metal plate, with 2 inches of external insulation. The cell is painted white and receives direct solar radiation all day. Inside the cell, 190 concrete blocks (190 mm × 90 mm × 55 mm) comprise the thermal mass of the cell.

The pond was built using four 3 ft × 9 ft (0.91 m × 2.74 m) metal frames connected with screws and L connections. The area of the pond is 81 ft^2 (7.52 m^2). Inside these frames are R11 (1.937 m^{2}K/W) 5 in. (127 mm) thick insulation panels. Two insulating panels are also floating over the water, which is held inside the insulation by a heavy-duty plastic liner so that there are no leaks. The initial water level of the pond was 18 inches (0.455 m), for a volume of 121 ft^3 (11.2 m^3). There are two independent water systems in the La Roche and Givoni (2000) experiment. The first one is the cooling system for the test cell, which pumps water from the pond to the roof of the experimental chamber and can run either continuously or at selected hours of the day using a timer. The other system is activated at night, cooling the pond by pumping water out of the tank and spraying it over the insulation.

Water circulates over the floating insulation panels at night, cooling by evaporation, convection, and radiation, and during the day these panels block solar gains to the water, keeping it cool. The pond is 6 m from the test cell, and the tubes that connect to the experimental chamber are 10 m long and are covered by insulation and aluminum foil to avoid overheating from the sun. The water circulates through this

FIGURE 7.26 Roof pond system with floating insulation at UCLA.

tube into the concrete roof of the cell, where it penetrates and cools from the inside, acting as a heat sink and absorbing heat from the building, keeping it cool during the daytime.

Several tests were performed and predictive equations were developed, but the equations demonstrated the direct relationship between the wet-bulb temperature and the performance of the system. One of the equations predicted the performance of the system as a function of the wet-bulb temperature and solar radiation after a 15-day cooling series.

The indoor average dry-bulb temperature was found to correlate best with the average wet-bulb temperature and the equation to determine maximum DBT inside the test chamber was

$$T_{mx} = WBT_{avg} + \Delta t \tag{7.5}$$

where
- T_{mx} = predicted maximum dry-bulb temperature inside the chamber on a particular day
- Δt = average elevation of the indoor maximum above the outdoor average
- WBT_{avg} = average wet-bulb temperature for the day

In this case $\Delta t = 4.4$ K; however, this will change depending on the performance of the roof pond. Solar radiation can be included as another factor to determine the indoor temperature. Computed value where plotted with measured values and close agreement was found (Figure 7.27).

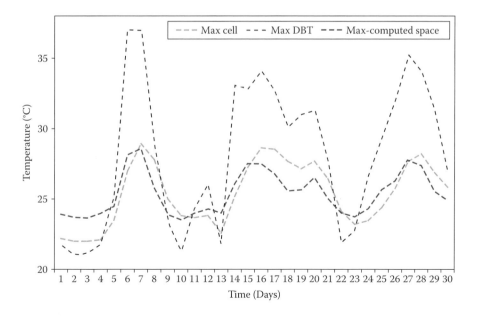

FIGURE 7.27 UCLA roof pond evaluated in a 30-day period.

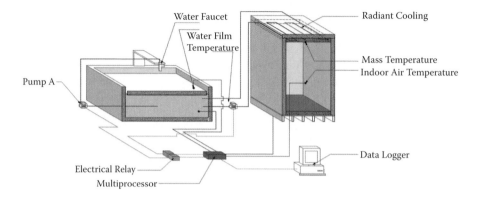

FIGURE 7.28 Roof pond system with floating insulation and smart controller at UCLA (summer daytime).

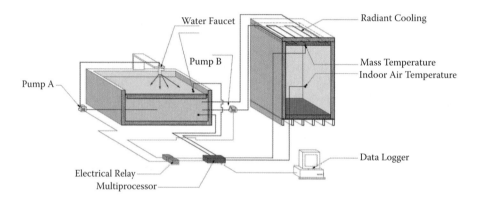

FIGURE 7.29 Roof pond system with floating insulation and smart controller at UCLA (summer nighttime).

Another series of tests was performed in which the roof pond pump was not controlled by a timer but instead had sensors that operated a system according to relationships between indoor and outdoor dry-bulb temperature. A datalogger was connected to several sensors that measured and compared temperature values. Rules were established that compared the temperature values in each of the sensors, and depending on the conditions, operated the pump by either turning it ON or OFF (Figures 7.28 and 7.29). New versions will operate the system comparing indoor temperature with outdoor wet-bulb temperature, which is a better indicator of pond performance.

Results from the tests with these roof ponds indicate that the indoor temperature in the cell without any cooling system activated (reference condition) is mostly a function of the outdoor average temperature; but when the building is evaporatively

Passive Cooling Systems

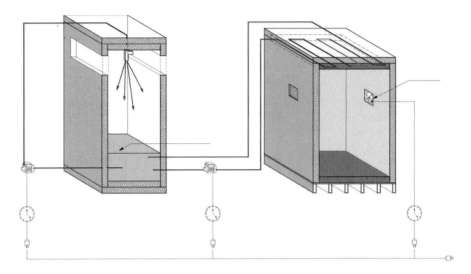

FIGURE 7.30 Diagram of shower system with indirect evaporative cooling.

cooled, the indoor temperature is mostly a function of the outdoor average wet-bulb temperature, with the performance of the system improving with lower wet-bulb temperatures and higher wet-bulb temperature depressions. Even when night ventilation is combined with evaporative cooling, the dominant variable in the determination of air temperature inside the cell is the wet-bulb temperature (and higher WBT depressions). Other factors such as solar radiation and daily temperature swing were explored, but the wet-bulb temperature was always the dominant factor to determine evaporative cooling potential, making it easier to determine by knowing this single variable. These tests demonstrated that the pond, which is the main component of the cooling system, could be separate from the building.

Another indirect evaporative cooling system with a small shower that was only 2.40 m high was built (Figures 7.23 and 7.30). As with the roof pond, the indoor temperature was also a function of the outdoor wet-bulb temperature.

7.4.2.2 Roof Ponds in a Hot and Humid Climate

Even though roof ponds are traditionally implemented in hot and dry climates, they have also been successfully used by Kruger, Gonzalez, and Givoni (2010) in a hot and humid climate in Maracaibo, Venezuela. Maracaibo has year-round maximum average temperatures between 32 and 34°C, with minimum relative humidity between 50 and 60% and a small daily swing of about 5 to 6°C. Gonzalez et al. (2000) tested several configurations in test cells, including some in which a small pool of water was open to the air but was shaded. The drop of the indoor maximum compared with the outdoor maximum was about 3°C, or half the daily swing.

Gonzalez, Machado, Rodriguez, Leon, Soto, and Almao (2000) then built a full-scale sustainable affordable house, also for Maracaibo. The roof pond in this house was shaded by a reflective horizontal plate, and air movement between the cover

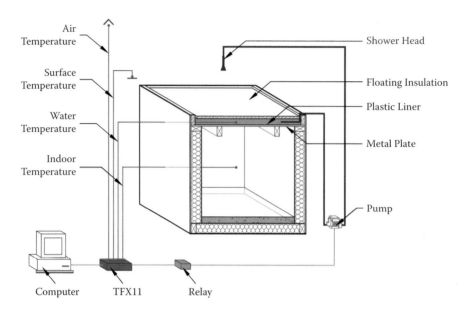

FIGURE 7.31 Smart roof pond at Cal Poly Pomona daytime.

and the water was provided through screened pipes in the side walls and by small exhaust fans. Equations were developed by Kruger et al. (2010) to predict the indoor maximum, average, and minimum temperatures. When the spaces were occupied, the indoor maximum temperature was about 2.2°C below the outdoor maximum.

7.4.2.3 Cal Poly Pomona Smart Roof Pond with Floating Insulation

A roof pond with floating insulation in which the night circulation of the pond water is controlled by a smart controller was developed and tested at the Lyle Center for Regenerative Studies at Cal Poly Pomona (La Roche, 2006). Initially the pumps were controlled with a timer, but then a second set of series was performed in summer 2006, with a smart controller operating a pump connected to a roof pond. Several cells were built that were 4 ft wide by 4 ft high with walls built with 2×4 studs and fiberglass insulation, plywood outside, and gypsum board on the inside. Both cells have a south-facing $2\,\text{ft} \times 2\,\text{ft}$ window for a glazing-to-floor ratio of 25%. All the test cells were lightweight, without thermal mass.

Several rules were tested, and the simplest one compared the temperature of a metal painted black surface facing the sky with the air temperature. If the temperature of this surface was higher than the air temperature, then the pump would be OFF (Figure 7.31). If the temperature of this surface was lower than the air temperature, then the surface was radiating to the night sky and the pump was activated. This would happen at night (Figure 7.32). A more complex rule was

If $T_{surf}(A) < Outdoortemp(A) + 2$ and $Roofpond(A) > Comfortlow$ and $T_{water}(A) >$ *lowatertemp* then PUMP ON (7.6)

Passive Cooling Systems

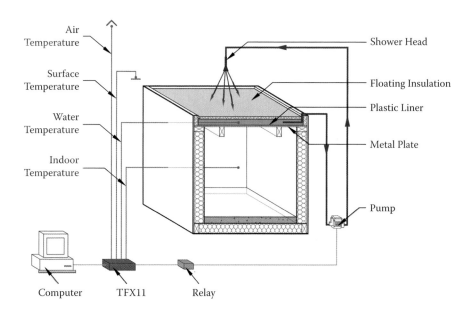

FIGURE 7.32 Smart roof pond at Cal Poly Pomona nighttime.

where

$T_{surf}(A)$	= Temperature of the external metal plate
$Outdoortemp(A)$	= Outdoor dry-bulb temperature
$Roofpond(A)$	= Temperature in the test cell with the roof pond
$Comfortlow$	= Lower limit of the comfort zone
$T_{water}(A)$	= Temperature of the water
$lowatertemp$	= Lowest minimum temperature of the water

Figure 7.33 shows indoor temperatures in the control test cell (without any cooling) and the cell cooled by the smart roof pond, with the timing of the pump operation. The cell with the roof pond is always the coolest one, about 8°C below the control cell. These results demonstrate the effectiveness of the system and that indoor daytime temperatures can be much lower than the outdoors even in lightweight buildings.

7.4.2.4 Cal Poly Pomona Modular Roof Pond

Marnich, Yamnitz, La Roche, and Carbonnier (2010) developed and tested a modular roof pond system at Cal Poly Pomona. The purpose of this modular roof pond system was to develop a modular roof that could be installed with relative ease above any new or existing roof (that would support the load) without affecting the integrity of the structure. Another goal was that it would have a minimum of moving parts to reduce maintenance of the system so that it would be easier to adopt in many projects.

Two types of 4 ft × 4 ft modules were tested in different configurations to determine cooling potential. The two main types of configurations included one

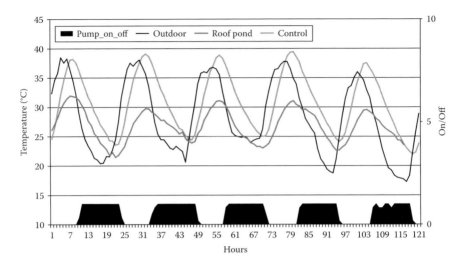

FIGURE 7.33 Indoor temperatures in the control and the cell cooled by the smart roof pond.

with bags full of water that cooled only by radiation and another with exposed water that had sprays of water that provided evaporative cooling. Both of these had fins that opened during the night and closed during the day. Some configurations had fixed fins in a position that blocked direct solar radiation during the daytime but permitted some sky exposure during the night for radiant cooling. All types of roof ponds produced lower temperatures than the control cell, but the rood pond with the water sprayed horizontally at night across the pond below the shading panels cooled more than any other condition (Figure 7.34). The performance of these systems can be further improved, for example, by adding more intelligence to the controller and operating the nozzles and maybe the fins of the roof pond at different times according to rules.

La Roche, along with several students, developed a roof pond with a system of operable louvers that work as a shading system and could be controlled by rules or a timer based on daytime or nighttime. If operating the louver system based on the time of day, it must be kept closed during the daytime and opened during the night (Figure 7.35). A sprinkler attached to a roof pond increases the amount of cooling by evaporation that takes place within a system. The students proposed to integrate these evaporative cooling systems with Cal Earth's Ecodomes, providing cooling in hot and dry climates (Figure 7.36).

7.4.2.5 University of Nevada, Las Vegas Roof Pond

Fernandez-Gonzalez and Hossain (2010) conducted promising research in the NEAT Lab at the University of Nevada, Las Vegas (UNLV) School of Architecture. Fernandez has continued to test variations of the Harold Hays Skytherm System in two identical 2.69 m² test cells, one of which has a 22.9 cm deep roof pond over its ceiling, which is made of corrugated metal deck. The

Passive Cooling Systems

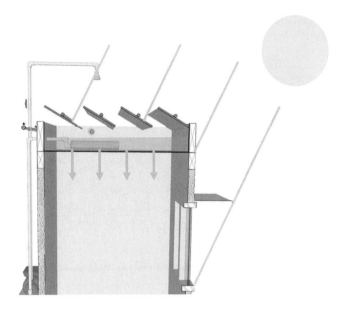

FIGURE 7.34 Cal Poly Pomona modular roof pond with fixed fins.

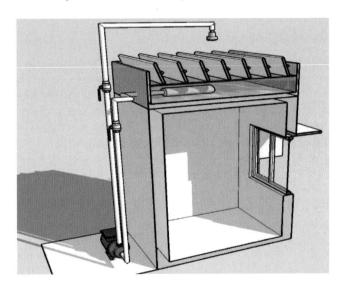

FIGURE 7.35 Operable louver roof pond at Cal Poly Pomona.

walls are R16, and the standard automated garage doors provide the movable insulation for the roof pond. Both cells were shaded during the daytime with the garage door, and the cell with the roof pond had maximum average temperatures 6.6°C below the summer average maximum outdoor air temperature. If a spray of water was added, then the temperature could be lowered even more.

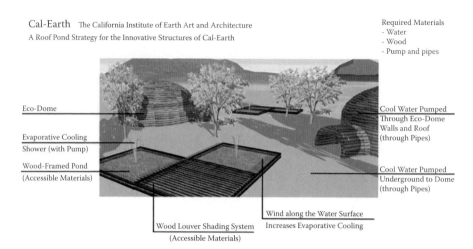

FIGURE 7.36 Evaporative system integrated with Cal Earth Ecodomes.

7.5 THE UPPER ATMOSPHERE AS A HEAT SINK: RADIANT COOLING

7.5.1 Principles of Radiant Cooling System

Radiation is essential in maintaining the Earth's thermal balance. Outgoing radiation releases energy to outer space from the Earth and its atmosphere, and cosmic space is the ultimate absorber, balancing the energy inputs from the sun as well as from other sources. The sky provides the ultimate continuous heat sink to maintain the Earth's thermal equilibrium. It is the one passive cooling mechanism that parallels the only means of passive heating, the sun's incoming radiation.

All bodies radiate and absorb energy to the sky at the same time, and usually at different rates and wavelengths. Any ordinary surface that "sees" the sky loses heat by the emission of longwave radiation toward the sky and is regarded as a heat radiator. These radiant heat losses take place during day and night, but only at night, when there are no solar gains, are the heat losses to the sky higher than the heat gains (negative radiant gains) so that the building can actually be cooled. Starting at sunset, the exterior surfaces of buildings and the surrounding environment begin to be cooled by net thermal losses in radiation. But the cooling power by sky radiation is quite small compared with the solar irradiation reaching the Earth. For example, in the southern United States, summer radiative cooling rates of 100–200 Btu/ft^2 (1,000–2,000 kJ/m^2) per night for a dry horizontal surface at 74°F (21°C) can be expected. This is similar to the radiant gains in 1 hour of daylight in the same location. Thus, a rule of thumb for a location with clear days and nights is that radiative cooling has a potential rate of about 10% of the summer radiative heating rate (Cook, 1989, p. 33).

One of the advantages of radiant cooling is that it has the least thermal impact on the immediate exterior environment. All that is needed is an exterior environment that does not impede the clear view of the sky: a roof facing the sky.

Five heat transfer processes affect the performance of any radiant cooling system (Givoni, 1994):

1. Longwave radiation emitted by the radiator; this is augmented by clear skies and reduced when overcast
2. Radiation emitted by the deep sky and absorbed by the radiating surface
3. Convective heat exchange between the radiator and the ambient air
4. Cold energy transfer from the radiator to the building, either by conduction or by forced airflow in the case of specialized lightweight metallic radiators
5. Convective heat exchange between the radiator and the heat transfer medium, the air (in the case of a lightweight radiator)

The system's performance depends in part on the climatic conditions and in part on the design details. The balance between the first two processes yields the net radiant heat loss, which is the climatic potential for radiant cooling. The balance between the net radiant loss and the convective exchange with the ambient air yields the stagnation temperature of the radiator. The heat exchange between the radiator and the air flowing underneath and the airflow rate determine the temperature of the air exiting from the radiator and the cooling energy delivered to the building.

The intensity of longwave radiation emitted by a specialized radiator or by the roof itself when it serves as the radiating surface depends on only two factors: (1) the absolute temperature of the radiating surface (T_r), and (2) its emissivity (E_r) as defined in the Stefan–Boltzmann law:

$$R_e = S \times E_r \times T_r^4$$

where

R_e = Intensity of longwave radiation emitted by a specialized radiator
S = Stefan–Boltzmann constant 0.567×10^{-8} (W/m² K⁴)

The emissivity of a radiating surface is a physical property that represents the potential of a surface to emit radiation relative to a perfect black surface. Most common construction materials, except metals, have emissivities close to 0.9. If a metallic material is the radiator, then it should be covered with paint, which usually has an emissivity close to 0.9. The paint layer then becomes the radiating surface.

The temperature of the radiator, and especially its relation to ambient air temperature, also affects the performance of the radiator. If the temperature of the radiator is higher than that of the air, then the radiator also loses heat by convection to the air.

The cloudiness of the sky affects the potential of heat losses to the upper atmosphere, because clouds emit radiation through the whole longwave spectrum. Under an overcast sky, radiant losses to the upper atmosphere are reduced, and the clouds literally block solar radiation losses to the sky.

One of the first recent radiant cooling systems was the Skytherm System developed by Harold Hay in 1978. In this system, the structural roof consists of a horizontal uninsulated metal deck, and thermal mass is provided by large plastic bags filled with water that are placed above the metal. Horizontal insulating panels, moved by a motor, insulate the water bags during the daytime and expose them to the sky at night, cooling the water bags by radiation. The metal ceiling serves as a radiant cooling panel for the space below. During the winter this process is reversed, and the system acts as a passive solar heating system in which the water is heated.

7.5.2 UCLA Radiant Cooling System

In developing countries, a large portion of the population lives in poverty, mostly in informal settlements with minimum access to basic living conditions. In these neighborhoods, most of the housing is generally built in stages using any available materials. A common initial stage of these dwellings is to have tin or zinc sheets for the walls and roofs (Figure 7.37) or brick walls with metal sheets directly above the walls. Usually the metal walls are replaced by brick walls as soon as the family has the financial resources to do this. These metal roofs, which are very common in many developing countries, are very hot during the daytime but cool down quickly in the evening as they radiate more energy to the night sky when they no longer receive heat from the sun.

In hot climates, dry-bulb temperatures during the day inside these dwellings are much higher than the outdoors, especially on sunny days, but values drop rapidly at nighttime until they are close to outdoor conditions.

In December 2000, data were recorded in a house with metal walls and roofs in Maracaibo, a city located in a hot and humid climate, at a latitude of 10.5 degrees north of the equator and at sea level. Even though it was cloudy during the days in which measurements were taken, indoor daytime temperatures were always much higher than outdoor temperatures. Two adults and three children live in the 40 m² house. Cooking is done inside the house, but dining and most of the daytime activities are done outside under an *enramada*, which is a traditional covered outdoor space, usually in the back of the house and open to the backyard. While the porch faces the street and connects the house with the neighborhood, the enramada is for more private outdoor living.

FIGURE 7.37 House in an informal settlement in Venezuela.

Passive Cooling Systems 261

Because of its proximity to the equator, daytime conditions inside the house would improve by adding insulation on the roof, but this would reduce the potential for radiant cooling during the night. Installing operable hinged interior insulation plates, even as simple aluminum foil, under the roof reduces daytime heating without interfering too much with the cooling effect of such roofs during the nights when the louvers would be open.

The effectiveness of storing the cold energy generated by the nocturnal radiant cooling for lowering the indoor air temperature during the daytime hours depends on the thermal storage of the interior space (Geros et al., 1999; Givoni, 1992). If mass is also present, as in the common brick walls, then it can be used as a heat sink during the daytime while the panels are closed to form an attic that acts as insulation.

The test cell used in this experiment was built by the Energy Lab in the roof of Department of Architecture at UCLA by Givoni, Gomez, and Gulish (1996). Its internal dimensions were 1.00 m × 1.00 m × 0.95 m, and the walls and floor were super-insulated with polyurethane panels 80 mm (3.5 in.) thick so that the radiant system is the main source of cooling. Many other series were conducted with different levels of insulation in the walls to further approximate typical buildings. The projected area of the gabled-sloped metal roof was 1.16 m × 1.60 m. The roof and the walls were painted white (Figure 7.38).

After painting the roof, it acted as a nocturnal radiator. Below it were located centrally hinged, lightweight, and operable reflecting panels. During the daytime the panels were in a horizontal position (closed), which form a continuous insulation

FIGURE 7.38   Smart operable radiant cooling system at UCLA.

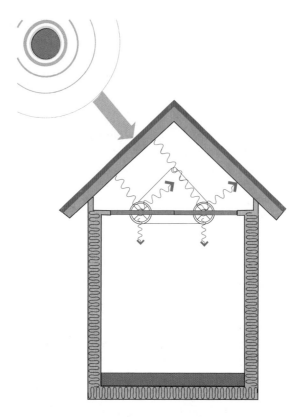

FIGURE 7.39 UCLA radiant cooling system during daytime.

under the roof and thus reduce the heat flow into the interior space (Figure 7.39). During the night, the panels were turned into a vertical position, enabling radiant and convective heat flow from the interior space to the metal ceiling, which is cooled by longwave radiation to the sky (Figure 7.40).

The rotation of the insulating ceiling panels between the horizontal (closed) and vertical (open) positions was achieved by an electromechanical system designed and built by Gomez. Under computer control, the panels were opened at sunset and closed at sunrise, according to a prescribed action schedule calculated for the longitude and latitude of Los Angeles. In a real home the panels could easily be controlled manually from the interior (e.g., by a rope). Interior operable insulation panels are not exposed to the wind and the rain, and thus can be simpler in construction, lighter, and much less expensive than external operable panels.

A series of experiments to determine the performance of the operable system with different combinations of mass were performed during several years between 2000 and 2002. Concrete bricks with dimensions of 89 mm × 57 mm × 190 mm were used as thermal mass. The position of these bricks is changed in the different series from the floor of the cell to the walls, simulating the behavior of a building with a floor slab or brick walls. The distribution of the mass inside the space affected the temperature distribution and performance of the system. When the mass was located

FIGURE 7.40 UCLA radiant cooling system during nighttime.

in the wall, the temperature was relatively uniformly distributed up to the height of the wall. On the contrary, when the mass was placed in the floor, most of the cool air stayed close to its surface, below a level in which it would be useful for occupants.

7.5.3 Zomeworks Double-Play System

Zomeworks has designed several radiant cooling and heating systems. *Double-play* is a system that uses overhead water storage as the working fluid for heat transfer and thermal mass. This system uses plain water as the heat transfer fluid and night sky radiation for cooling. The water circulates between tanks in the occupied rooms and absorber radiator arrays on the outside of the buildings. A solar thermal loop is used for passive heating. Results of testing over 2 years have been presented, and results are very promising (Beauchamp 2010). In the summer the system removes heat from the building at a rate of about 4,875 Btu/hr or 39,000 Btus in an 8-hour period, reducing the temperature of an 800 gallon tank 6°F. In the winter the system absorbs 52,000 Btu in a period of 7 hours, raising the temperature of the same 800 gallons by 8°F.

7.6 THE EARTH AS A HEAT SINK: EARTH COUPLING

Ground cooling occurs when heat is dissipated from the building to the ground, which during the cooling period has a temperature lower than the ambient air temperature. This dissipation can be achieved by direct contact of an important part of the building envelope with the ground or by injecting air that has been circulated underground into the building, using earth-to-air heat exchangers.

The temperature of the Earth is relatively stable at depths between 6 and 30 meters. The long-term annual average is usually about one degree above the mean annual air temperature and can be estimated by measuring the temperature of local well water. As we approach the surface, the Earth's temperatures track the air temperature more closely, and the ground surface temperature follows the sinusoidal pattern of solar heat flux. In a mild climate with cool winters, during the summer, the Earth's temperature below a depth of two meters will probably be lower than ambient air, and the heat that a building gains from internal and solar gains can be transferred by conduction to the ground that surrounds it. If this ground temperature at a depth of 2–3 m is around or below 22°C, the soil acts as a heat sink for buildings. If the temperature of the Earth is above 22°C, it is still possible to use the ground as a cooling surface by cooling the surface of the soil below its average temperature through shade that blocks solar radiation from the sun, and evaporative cooling at the soil's subsurface. The surface ground temperature amplitude is the maximum departure of temperature from the annual mean. If there were no other variables in play, it would increase, as solar radiation, with latitude.

Two additional considerations when designing an underground system are control of condensation and ventilation. It is also important to balance winter versus summer conditions. Summer losses would be beneficial; however, winter losses would not.

7.6.1 Ground Cooling of the Building by Direct Contact

A building exchanges heat with the environment by convection, radiation, and conduction. Conduction to the ground is usually through the slab. The conductive heat exchange between the building and the ground can be increased by maximizing the area of the building in contact with the ground (Figure 7.41). Because conductive

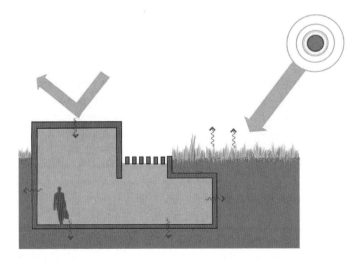

FIGURE 7.41 Cooling a building by direct earth contact.

exchanges between the building must be augmented to increase heat losses from the building to the Earth, it is not possible to use uninsulated walls in an earth-sheltered building located in a climate with cold winters because all of its heat will be lost to the Earth. If the building must be heated, then its walls should be insulated to avoid heat loss to the thermal mass around it.

Buildings can be earth-covered to varying degrees. A bermed structure is built on the surface grade with earth added around it and can be constructed below grade or into hillsides and be totally or partially covered by earth, including the walls and parts of the roof. As the surface of the building in contact with the thermal mass of the earth around it increases, its thermal inertia also increases, reducing temperature swings.

7.6.2 Ground Cooling of the Building by Earth-to-Air Heat Exchangers

If the building does not have enough surface in contact with the ground, then cooling can be supplemented by using earth-to-air heat exchangers, indirectly increasing the amount of earth in contact with the building. These exchangers are pipes, or *earth tubes*, which are buried horizontally at about 2 m deep, through which the air circulates by means of electric fans that suction it from the outside. As the air circulates through the outlet, it loses its heat to the earth around it, and during the summer, when the ground temperature is lower than the outside air, the air temperature at the outlet will be lower than the temperature at the inlet. Because it is important to transfer heat between the pipe and the earth, the material should be conductive, and plastic, concrete, or metallic pipes are used. The temperature drop depends on the inlet temperature, the air velocity and volume of air through the pipe, the ground temperature at the depth of the heat exchanger, the thermal conductivity of the pipes, and the thermal diffusivity of the soil. Moisture in pipes can also lead to mold and mildew problems. Because of the air's low heat capacity and the relatively high temperatures and low air velocities that are obtained with earth tubes reducing their cooling potential, these are usually not competitive with mechanical air conditioning solely on an economic basis.

Earth tube systems are usually classified as open- or closed-loop systems. In an open-loop system the air is drawn from the outdoors and delivered to the interior of the building, providing ventilation and cooling. The air that enters the building collects heat from the building and then exits to the atmosphere around it (Figure 7.42). In a closed-loop system the air circulates through the tubes and then back again. A typical closed-loop system would not connect the heat transfer fluid with the space; however, because air is used as the mechanism and it has such a low heat capacity, it is better to include the building space as part of the loop. The cool air enters the building from one side and exits through the other side to be cooled and reused again (Figure 7.43). Because this air coming out of the building might still be cooler and less humid than the outside air, a closed-loop system can be more efficient than an open-loop system.

It is usually assumed that earth tubes can provide sufficient cooling when ground temperatures are below 13°C. Because the Earth's temperature is about the same as the yearly average temperature, this will occur in mild to cold climates but not in a climate that is warm during the whole year, especially in a hot and humid climate in

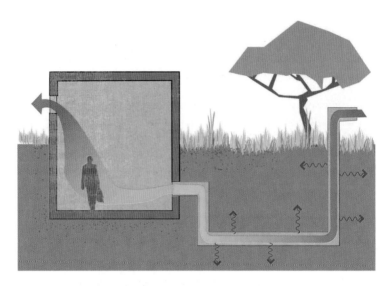

FIGURE 7.42 Cooling a building using open-loop air exchangers.

FIGURE 7.43 Cooling a building using closed-loop air exchangers.

which average yearly dry-bulb temperatures are above 21°C. Even though the air can be cooled underground, it will not be cool or dry enough to provide sufficient cooling inside the space. However, a report by Sharan and Jadvah (2003) demonstrates successful use of earth cooling with pipes with average earth temperatures of 27°C and different comfort standards. With a soil temperature of 26.6°C and an air temperature of 40.8°C, an outlet temperature of 26.6°C was produced in a 50 m pipe that was 10 cm in diameter, 3 m below the surface, and with an air velocity of 11 m/s. This demonstrates how, because of its low heat capacity, the air temperature in the tube reaches the ground temperature quickly. Tubes need not be very long or very large in diameter; 20 to 30 cm is a recommended diameter. They should be placed at

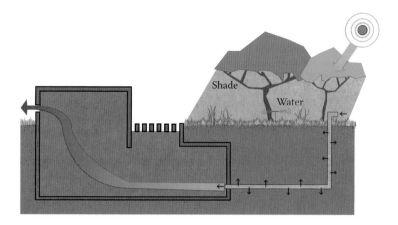

FIGURE 7.44 Cooling the earth around a building.

least 2 m below the ground; and if the soil has moisture or is cooled above ground, it will perform better. The tube should be conductive to transfer the energy from the air to the earth, so plastic or metal perform about the same.

7.6.3 Cooling the Earth

In temperate climates the earth mass under and around the building is a good cooling source for earth tubes, but in warmer climates the temperature of this mass might be too high. It is possible to use this mass of earth by reducing its temperature below that of the exposed soil. This can be done by either surface cooling or subsurface cooling (Figure 7.44).

Surface cooling is lowering the temperature of the soil surface and consequently the temperature of the earth's mass below the surface, providing a large amount of cooled mass. The simplest way to cool a surface is by shading to prevent the soil from being heated. This can easily lower the surface temperature by 8°C to 10°C (14°F to 18°F). Another strategy is to cover the soil with wood or a gravel mulching layer at least 10 cm thick that can be irrigated at certain hours. Solar radiation is absorbed and intercepted within the top 3 cm or 5 cm at the top of the layer, and the conduction of the heat down the layers of pebbles is rather low. The moisture from the irrigation is evaporated and the energy to do this comes mostly from the soil, lowering its temperature. Also, if the building is raised above the ground, the soil underneath can be humidified so that its evaporation will cool the air below the house.

7.7 APPLICABILITY OF PASSIVE COOLING SYSTEMS

The main features of passive cooling systems are presented in Table 7.3. This table includes the natural heat sink used to transfer the heat from the building, climatic applicability that describes outdoor conditions under which it would perform appropriately, physiological effect, and the dominant building element that will affect its performance.

TABLE 7.3
Passive Cooling Systems

Passive Cooling System	Heat Sink	Climatic Range	Physiological Effect	Dominant Building Element
Comfort ventilation	Air	Max temp < 30°C RH < 80% WBT < 24°C	Increases sweat evaporation from the skin. Cooling and convective heat loss from the body.	Large openings to promote air movement Low mass to avoid storing heat
Nocturnal ventilative cooling	Air	Large diurnal range >12°K Min DBT < 20°C Max DBT< 38°C RH Absolute humidity < 14 g/kg	Decreases MRT of interior walls	High mass with external insulation Windows should be operable to permit air exchanges at night but closing at day
Radiant cooling	Space or upper atmosphere	Clear nights Humidity and temperature are not as important	Reduces MRT of interior walls	Roof surface to the sky
Evaporative cooling: Direct and indirect	Air. Uses water as transport (direct) and heat sink (indirect)	WBT< 21°C DBT < 42°C WBT< 24°C DBT < 44°C	If RH is low, it can be raised increasing comfort. If RH is high, an indirect system would be better.	Roof design or tower. Could also be features in the landscape around a building
Earth coupling	Earth	Average yearly DBT < 13°C for earth tubing For direct coupling if climate is temperate, soil temperature below 2 m will probably be cooler than air temperature If outdoor temperature is hot, surface temperature above the ground can be cooled to reduce underground temperature	Reduces air temperature and increases comfort by increasing the effect of cooling sensation by radiation	Relationship to earth, either sideways, above or below

Note: RH: Relative Humidity; WBT: Wet Bulb Temperature; DBT: Dry Bulb Temprature; Min DBT: Minimum Dry Bulb Temperature; MaxDBT: Maximum Dry Bulb Temperature.

8 Passive Heating

8.1 APPLICABILITY OF PASSIVE HEATING

In the psychrometric chart, if environmental conditions are to the left of the comfort zone, then the body loses more heat than it gains, and heat must be provided to achieve thermal comfort. Passive solar heating provides this heat to the body by warming the space around it. In general, conditions to the left of the comfort zone can benefit from passive solar heating if there is enough solar radiation to provide the necessary energy and if the building has the appropriate design features to capture energy, thermal mass to store it, and insulation to keep the heat inside. When outdoor conditions are below but still relatively close to the comfort zone, thermal mass or internal gains are sufficient to achieve thermal comfort, and requirements are more lenient. But as the outdoor temperatures become lower, the building must be designed with more stringent requirements (e.g., better insulation, tighter and better windows, additional thermal mass). It is generally assumed that the lowest outdoor temperatures with which we can provide sufficient heating to a building, using only passive solar strategies, is around 5°C (Figure 8.1). However, there are examples such as the Rocky Mountain Institute in which the building uses special glazing, super insulation, and large amounts of thermal mass to achieve thermal comfort with outdoor temperatures well below 5°C.

8.2 CONTROL OF HEAT LOSS

In cold climates, a building will lose heat by conduction through its walls, roof, and floor; by convection (infiltration) through the window frames; and by radiation through sky-facing surfaces (Figure 8.2). The building can also gain heat by radiation when the sun is available, although this is not considered when sizing mechanical systems.

The first step to achieve thermal comfort inside a building in the winter is to reduce heat losses from the interior of the building to the colder exterior. These strategies were discussed in Chapter 6 and include the reduction of heat losses by conduction and infiltration. Losses by conduction can be reduced by improving the insulation of the building skin, creating a more compact building, and making a tighter envelope. Infiltration losses can occur through all the cracks in the building, including the bottom of the drywall, edges of the window, and even the plumbing and electrical fixture conduits. These can be reduced by including an air infiltration barrier system or moisture barrier, adding weather stripping, and using caulk to fill holes in siding or masonry or to seal the transition gap between two different types of finishes. Infiltration rate should not be reduced below a value that compromises

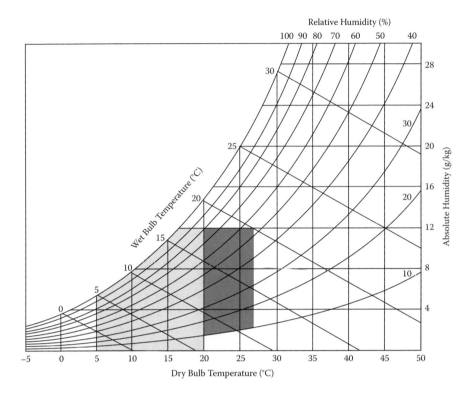

FIGURE 8.1 Outdoor conditions for which passive heating strategies will help achieve thermal comfort.

the indoor quality of the space. Emissions from occupants (CO_2) and construction materials inside the building require air exchange with the outdoors to maintain appropriate indoor air quality. Reducing building heat losses by infiltration and conduction are not passive heating strategies, but they are necessary because they help to keep inside the building the heat provided by passive, active, or mechanical systems inside the building. This chapter introduces the most common passive strategies that can be used to heat a building.

8.3 PASSIVE SOLAR HEATING

The simplest method to passively heat a building is to increase solar gains to the interior of the building. Passive solar systems must have a medium to collect or capture solar gains, a medium to store this energy, a medium to distribute this energy, a well-insulated and well-sealed skin to keep this energy inside during the underheated period, and control elements to reduce heat gains in the overheated period (Figure 8.3).

A passive solar system will capture heat during the day coming from solar radiation and will store it for night use. To achieve this, it must have mass on the inside for thermal storage and insulation on the outside. The thermal mass must face the interior so that it can be heated from the inside and then emit this heat back to

Passive Heating

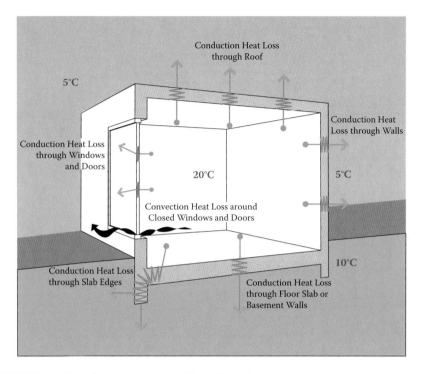

FIGURE 8.2 Heat flows through a building skin in the winter.

the occupants, and the insulation must be on the outside to keep the heat from the thermal mass inside during the night. When designing a passive solar system it is important to match the time when the sun can provide solar radiation to heat the space with the amount of thermal mass needed to store and provide the heat when it is needed inside the building. In general, a passive heating system will raise the average temperature of the building above the average temperature of the exterior and will have a smaller swing (Figure 8.4).

As we move farther away from equatorial latitudes, the optimum orientation for passive solar buildings is with the collection surface facing toward the equator because this is the facade that receives more solar radiation in the winter months. As the latitude increases, it becomes more and more important to have the collector surfaces facing south. It is usually considered good practice to keep the collector's orientation within 15 degrees of equatorial orientation with an unobstructed solar radiation window between 9:00 a.m. and 3:00 p.m. However, this varies as a function of specific microclimatic conditions such as morning or afternoon cloudiness. Generally, buildings oriented along an east–west axis provide more winter heating because they allow for maximum solar collection to the south during the whole day. This orientation is also advantageous during the summer because it minimizes east–west exposure, which is not helpful in the summer and is more difficult to control. Equatorial-facing collectors need not be all located along the same wall. For example, clerestory windows in buildings located in the northern hemisphere can project south sun deeper into the back of the building.

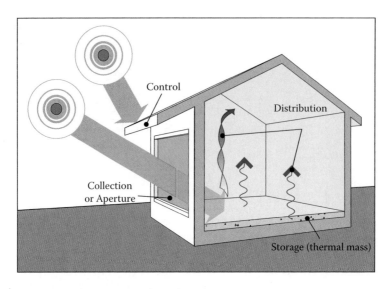

FIGURE 8.3 Components of a passive solar system.

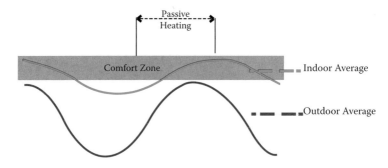

FIGURE 8.4 Effect of thermal mass and passive heating on indoor daily temperature swing.

In an east–west–oriented building, a polar-facing exterior wall will receive no direct solar radiation during the winter and becomes major source of heat loss. Because windows have a much lower U-value than walls, window surfaces on north-facing walls in cold climates should be small to reduce heat losses. Additionally, building shading of open spaces to the north of the building usually render them unusable for outdoor use during the winter.

Habitable spaces that are occupied the most with the highest heating and lighting requirement should be arranged along the equatorial facade of the building. However, overheating and glare can occur whenever sunlight penetrates directly into a building, and their control must be addressed through proper design. Rooms that are used the least (e.g., in residential buildings, closets, storage areas, garages) should be placed along the polar-facing wall where they can act as a buffer between high-use spaces and the exterior, which does not receive any radiation. An open floor plan optimizes passive system operation.

Passive Heating

A passive solar building that uses solar radiation as a heating source should also be designed to take advantage of solar radiation as a lighting source. However, each of these has different design requirements that must be addressed. In general, passive solar heating requires direct solar radiation to strike a surface and heat it, whereas daylighting requires diffusion of this radiation over larger areas of light-colored surfaces.

8.4 TYPES OF PASSIVE HEATING SYSTEMS

Passive heating systems are usually classified in three main groups: direct gain, indirect gain, and isolated gain systems. Direct gain systems use the actual living space as the solar collector, heat absorber, and distribution system. Indirect gain systems have the thermal mass located between the sun and the living spaces, and isolated gain systems have their components separated from the main usable areas.

8.4.1 Direct Gain Systems

The simplest passive solar heating system is a direct gain system that is an insulated building with internal thermal mass and equatorial-facing windows. In a direct gain system, solar radiation enters the living space and hits the thermal storage mass where it is stored to be reemitted later at night (Figures 8.5 and 8.6). Thus, the living space where the mass has been stored is a "live-in" collector. A direct gain system uses 60% to 75% of the solar energy striking the windows of the system.

The requirements for direct gain systems are a large amount of equatorial-facing glazing and exposed thermal mass in the floor, walls, or ceilings that are sized with enough capacity for thermal storage and positioned for solar exposure. The equatorial-facing glass admits solar energy into the house, where it falls on the thermal mass of the house. These materials store the energy during the daytime

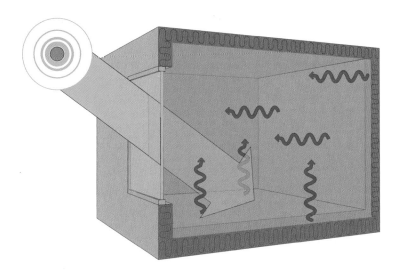

FIGURE 8.5 Direct gain system during the day.

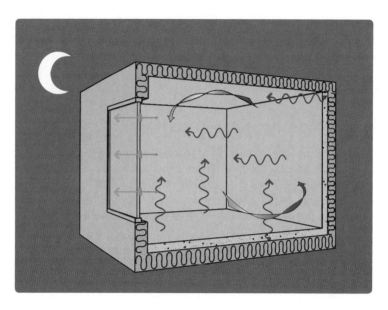

FIGURE 8.6 Direct gain system during the night.

and reradiate it to the space sometime during the night after the space is no longer receiving any solar radiation. Double glazing is used to reduce the losses at night, and operable insulation in the form of blinds can be used to increase the insulation level of the windows and to reduce heat losses by conduction.

Direct gain systems have several advantages. First, they are very simple to conceptualize and build. It is possible to create a direct gain system by simply placing the windows in the right location, and because they are simple the whole system is also inexpensive. Second, the large areas of glazing admit not only solar radiation but also natural light and provide visual connections with the outside. And third, glazing is readily available, and even the simplest most inexpensive windows can provide a greenhouse effect.

Direct gain systems also have several disadvantages. The large areas of glazing can create glare during the day and loss of privacy at night. In addition, ultraviolet radiation from the sunlight can degrade fabrics and photographs. Also, large glazed surfaces and large amounts of mass are needed for the system to be effective, and this additional mass can be expensive if it does not serve any structural purpose. And even with mass, diurnal temperature swings of 10 K are common. Additional nighttime insulation in the windows is also needed at night when outdoor night temperatures are very low.

The three most important factors to consider in the design of a direct gain system are the amount and orientation of the glazing, the amount and distribution of the thermal mass including its thickness and position, and the insulation level of the envelope. The glazing area determines how much solar heat can be collected in the building, and the amount of thermal mass determines how much of that heat can be stored. Insulation keeps the heat inside the space so that the energy stored by the mass is not lost to the outdoors. If the surface of equatorial-facing glazing is too small, then there will not be enough solar radiation penetrating into the space to heat it, and if there is too much glazing, there will be more losses than gains because of night losses. If there is not enough thermal mass, the space will have a large daily temperature swing, and it will be too hot during the day and too cold at night. If there is too much mass, some of it will be unused, but it will not affect the performance of the system, even though it could increase the load of a mechanical heating system.

There are some rules of thumb, but the best method to determine the required amount of glazing and thermal mass is to use energy modeling tools. The following two ratios of thermal mass to glazing are very rough rules of thumb for direct gain systems. Assuming the building has enough thermal storage mass in cold climates with average winter temperatures between –5°C and 0°C, there should be 0.2 m^2 to 0.45 m^2 of equatorial-facing glass for each square meter of interior floor area. In moderate climates with average winter temperatures between 5°C and 10°C, there should be 0.10 m^2 to 0.25 m^2 of equatorial-facing glass for each square meter of interior floor area. If night insulation is combined with glazing, these numbers can be higher.

As much thermal mass as possible should be placed in direct contact with the solar radiation inside the space. A larger surface area of the storage medium to glazing area will provide better heating, with a higher minimum air temperature and smaller temperature swing. For a 1:5 ratio of storage to glazing area, the fluctuation is 22°C, for 1:3 it is 14°C, and for a 1:9 it is 7°C (Balcomb, 1992). For thermal mass exposed only to air and not to direct solar radiation, about four times as much thermal mass is required than if it was in direct contact. Furthermore, if the thermal mass is in direct sunlight, the temperature swing of the room air will be about half of the swing of the storage. However, if it is not in direct contact with solar radiation, the temperature swing of the air inside the space will be about twice that of the storage, so four times more. It is also important to remember that masonry surfaces are less effective on a daily basis beyond a depth of 10 mm, as explained in Chapter 6. This means that it is usually better to have the same amount of thermal mass distributed and exposed over a wider surface to capture more solar radiation than concentrated in a smaller area with more thickness.

Storage walls should not be set back more than 1.5 times the height of the collecting surface because if they are farther back than this distance they will probably not receive direct solar radiation. This means that an interior thermal storage wall should not be farther than 4.5 m away from a 3 m tall solar-collecting glazing wall.

8.4.2 Indirect Gain Systems

In an indirect gain system the thermal storage materials are placed between the interior habitable space and the sun. These systems integrate collection, storage, and distribution functions within some part of the building envelope that encloses the living spaces. There is no direct heating from the sun to the space; the heating is provided by the element that has thermal mass to absorb the incident solar radiation that strikes it and then transfers it to the living space by conduction. Depending on the

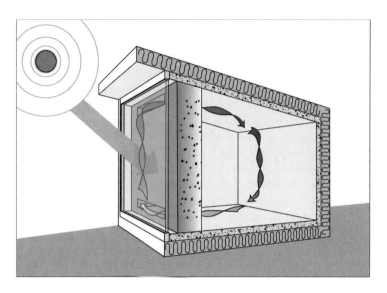

FIGURE 8.7 Trombe wall during the day.

thickness of the material this process can begin almost instantaneously or take many hours. This thermal mass is usually dark colored and has a glazed surface in the exterior face to create a small greenhouse effect reducing heat losses to the outdoors.

An indirect gain system uses between 30% and 45% of the solar energy that strikes the glass adjoining the thermal mass. The most common indirect gain systems are Trombe walls, mass walls, water walls, and roof ponds.

A Trombe wall has a glazed surface toward the exterior of the thermal mass separated by between 10 and 15 cm or more for maintenance. Solar radiation passes through this glazed surface to the mass wall, hits it, and is absorbed by it, causing the surface of the masonry to warm up. The glazed surface on the exterior keeps the longwave radiation from being lost to the outside (Figure 8.7). This heat is transferred to the inner surface of the wall by conduction, from where it radiates and is convected to the interior space. The time lag and dampening of the wall depend on the type and thickness of material; a rough guide is about 45 minutes per inch of material. Trombe walls also have vents located toward the top and the bottom that allow for the distribution of the collected air (which can reach 60°C) by natural convection and thus can provide immediate heating inside the space. These vents can be closed at night to prevent reverse circulation (Figure 8.8). Typically, the openings should be between 1% and 3% of the wall area and the distance between the bottom and top vent should be at least 1.8 m.

In a mass wall system, the thermal storage mass for the building is also a south-facing wall of masonry or concrete construction with the external surface glazed to reduce heat losses to the outside. The difference between a mass and a Trombe wall is that the Trombe wall has vents at the top and bottom to allow the air to circulate through the interior space, and the mass wall transmits the heat only by conduction to the interior surface and then radiation to the space. A cool mass wall is heated by the solar radiation during the morning (Figure 8.9), begins to heat up and transmit

Passive Heating

FIGURE 8.8 Trombe wall during the night.

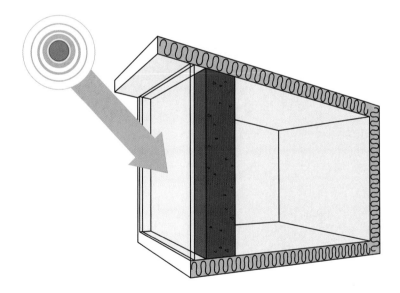

FIGURE 8.9 Mass wall early morning.

heat to the interior of the building during the day (Figure 8.10), and continues transmitting the heat collected from the solar radiation to the interior during the night (Figure 8.11) until heat inside the wall is depleted.

As with direct gain systems it is difficult to give sizing rules of thumb because of the number of variables involved. However several concepts are important to consider in the design of mass and Trombe walls: (1) Mass and Trombe walls should be between 15 cm and 30 cm thick for optimum performance. With even thicker

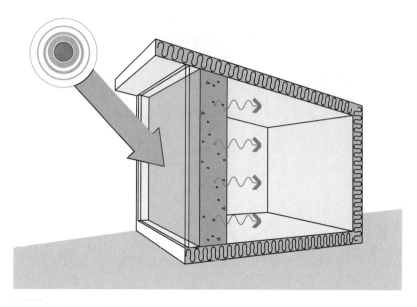

FIGURE 8.10 Mass wall midday.

FIGURE 8.11 Mass wall night.

walls (around 60 cm), thermal fluctuations inside the space are almost eliminated. (2) Different thicknesses also provide different time delays between the hot sunny exterior and the interior of the wall. Table 8.1 shows the temperature swing inside the space and the time delay between the irradiation in a double-glazed concrete wall on a sunny day and the occurrence of the peak temperature inside the space that

TABLE 8.1
Thermal Lag and Indoor Temperature Swing in a Concrete Mass Wall with Exterior Double Glazing on a Sunny Day

Wall Thickness cm (in.)	Indoor Temperature Swing °C (°F)	Time of Peak Temperature Interior of the Space
20 cm (8 in.)	15°C (27°F)	6 p.m.
9.5 cm (12 in.)	7.2°C (13°F)	8 p.m.
40.5 cm (16 in.)	3.6°C (6.5°F)	10:30 p.m.
51 cm (20 in.)	1.7°C (3°F)	1:30 a.m.
61 cm (24 in.)	0.7°C (1.3°F)	4:30 a.m.

Source: Anderson, B. *Passive Solar Design Handbook*, Volume 1, U.S. Department of Energy, 1981.

it encloses. (3) In the Trombe wall, the space can be heated by direct gain through the wall and convective loops through the vents during the day, and then through the thermal storage wall during the night. In mass walls with no vents there are no convective loops.

Mass and Trombe walls have several advantages. Glare, privacy, and ultraviolet degradation of fabrics are not a problem. Also, temperature swings in the living space are lower than with direct gain systems. And the time delay between absorption of the solar energy and delivery to the living space can be useful to better integrate to occupancy patterns and the outdoor requirements. There are several disadvantages, though. The external surface of the mass wall is relatively hot and connected to the external climate, leading to considerable heat losses. In addition, two south walls are required—one glazed and one mass wall. Also, discomfort can be caused at either end of the heating season by overheated air or by radiation from the inside surface on warm evenings. These can be controlled by venting. There is no view or daylighting with mass or Trombe walls, and the glazed wall must have access for cleaning. Also, water can condense on the glass, reducing its performance. In northern latitudes in the winter, the mass can be a heating burden on very cold or cloudy days.

A water wall is another type of indirect gain system. It is very similar to the mass and Trombe wall systems except that water replaces the solid wall. Water works better as thermal mass because water has a greater heat capacity per unit volume than brick or concrete and because the convection currents within the water cause it to act like an almost isothermal heat storage system. Furthermore, because water is a fluid it can be moved from one location to another, transferring this heat. During the daytime the water wall absorbs and almost immediately transmits energy to the interior (Figure 8.12), and then during the night the energy inside the wall continues to be transmitted to the interior of the space (Figure 8.13). A very rough rule of thumb for the design of water walls is to provide between 0.15 m^3 and 0.3 m^3 of water for each square meter of solar window. This is assuming that the interior water walls are dark colored and the wall receives direct solar radiation between the hours of 9:00 a.m. and 3:00 p.m.

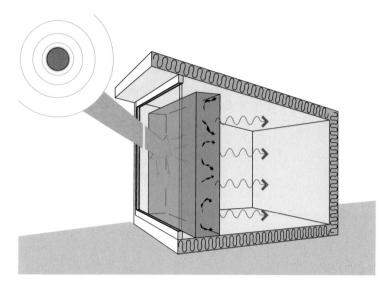

FIGURE 8.12 Water wall daytime.

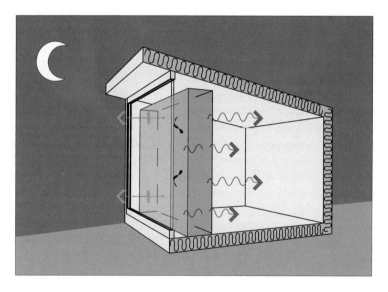

FIGURE 8.13 Water wall night.

Water walls have several advantages. The heat storage is isothermal, creating a lower external surface temperature and generating a lower temperature difference between the indoor and the outdoor and losing less energy to the outside. Also, glare, privacy, and ultraviolet degradation of fabrics are not a problem. Increased heat storage capacity means that the temperature swings in the living space are lower than with direct gain systems. And thermal storage can continue to supply energy to the living space well into the evening. Their disadvantages are first that water is difficult

Passive Heating

to contain and humidity can be a problem. Also, two south walls, one glazed and one mass, are required. Some water walls are in opaque containers, so there is no view or daylight. In addition, energy is lost to the outside air from the warm wall, and insulation can be expensive and awkward. If insufficient energy is provided during the day, as in a cloudy cold day, then the water wall will not be able to provide heat during the night and will be a heating burden for the mechanical systems. Another limitation for water walls and all wall indirect systems is the depth of the space determined by the height-to-depth ratio, a depth of about 8 m with typical wall heights.

A roof pond is an indirect gain heating system that can work as a radiant system, an evaporative cooling system, or both. A roof pond can be used for both heating and cooling. In a roof pond, thermal mass, generally in the form of enclosed waterbeds is placed horizontally, above an uninsulated ceiling of a building. In the winter, during the daytime this mass is exposed to direct solar gain, heats up, and transmits this heat by conduction to the ceiling, below which will then transmit this heat by radiation and some convection to the space below (Figure 8.14). At night and when the sky is overcast, insulation covers the warmed water and reduces the heat loss to the exterior (Figure 8.15) so that the roof pond transfers the stored heat to the interior.

Table 8.2 gives the ratios of roof pond collecting surface for each square unit of floor area in the interior space. Of course this will vary depending on location, exposure, and local conditions. The lower ratio given in Table 8.2 should be adequate at lower latitudes, while the higher ratio should be used at higher latitudes with colder climates. For latitudes higher than 36 degrees, roof ponds require greater exposure to solar gains as well as additional requirements to reduce heat losses. Roof ponds can also be tilted toward the equator in higher latitudes to maximize winter heating.

An isolated gain system has its components separate from the main living area of the building that is heated. This ability to isolate the system from the primary building areas is what differentiates this system from other systems. There are two main

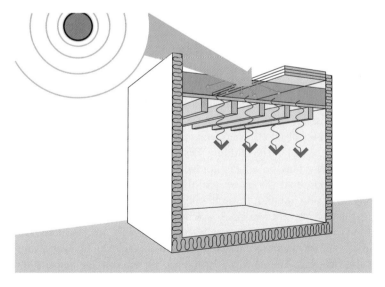

FIGURE 8.14 Roof pond during the winter day.

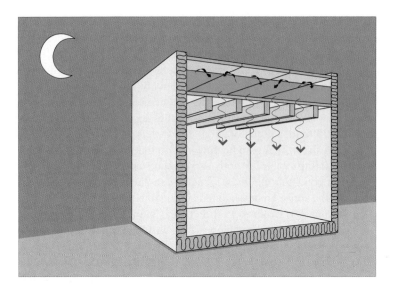

FIGURE 8.15 Roof pond during the winter night.

TABLE 8.2
Ratios of Roof Pond Collecting Surface per Unit of Floor Area (for Heating)

Average Winter Outdoor Temperature (°C)	−10 to −4	−4 to 2	2 to 8
Double-glazed ponds with night insulation	Not feasible	0.85–1.0	0.60–0.90
Single-glazed ponds with night insulation and reflector	Not feasible	Not feasible	0.33–0.60
Double-glazed pond with night insulation	feasible	0.5–1.0	0.25–0.45
South-sloping collector cover with night insulation	0.60–1	0.40–0.60	0.20–0.40

types of isolated gain systems: sunrooms and convective loops through an air collector to a storage system in the house. An isolated gain system will use 15%–30% of the sunlight striking the glazing to heat the adjoining living areas. Solar energy is also retained in the sunroom itself.

Sunrooms consist of an attached sunspace with a glazed enclosure built on the equatorial-facing facade of the building. Depending on the climate and how the sunspace is used, there may be a heat storage wall separating the sunspace from the building or an additional storage system that stabilizes the temperature. Vents and shading can also help to control overheating.

During the daytime, the sunspace is heated by solar radiation. Some of this heat is conducted through a thermal storage wall separating the interior of the house with the sunspace, but is transferred mostly by convection with small fans to the interior space of the building. The sunspace works as a heat collector and a solarium for people and plants (Figure 8.16). During the night the convection flows are blocked by closing the dampers in the vents, reducing heat transfer between the sunspace and

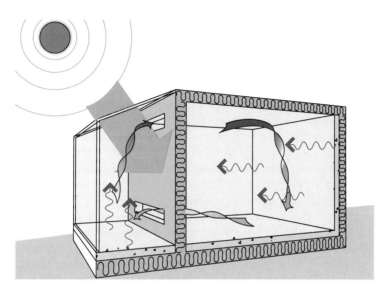

FIGURE 8.16 Isolated gain system day.

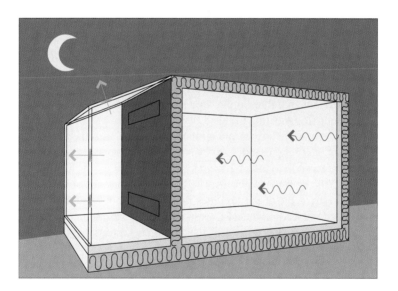

FIGURE 8.17 Isolated gain system night.

the interior spaces (Figure 8.17). Generally, in cold climates, a sunspace design will use between 0.65 m² and 1.5 m² of a south-facing double-glazed collecting surface for each square meter of floor area in the adjacent living space. In more moderate climates this number can be reduced to between 0.3 m² and 0.9 m² per square meter of floor area. The main difficulty with attached sunspaces is that they gain and lose heat rapidly, overheating during the day because they must be warmed up substantially to provide heat to the adjacent space, and then they overcool rapidly at night because they quickly lose heat through the glazing. This makes it difficult to use these spaces for a significant amount of time during the day or night.

8.5 EFFECTS OF DESIGN STRATEGIES ON EMISSIONS

Many different types of design strategies can affect emissions in different ways. For example, installing overhangs, adding insulation to the walls, using better windows, or implementing passive solar strategies will reduce emissions from building energy; living close to work or working at home will reduce emissions from transportation; and recycling will reduce emissions from waste decomposition and fabrication of materials. Some energy-related strategies have been compared to illustrate their effect on a building's greenhouse gas (GHG) emissions. The effect of these design strategies on emissions is studied in four climates. The strategies are (1) the insulation level of the wall, (2) the characteristics of the windows, and the efficiency of the (3) air-conditioning and (4) heating systems; the climates are hot and dry, hot and humid, cold, and temperate. The same single-family house with the same building emissions data was studied in each of the four climates with three levels of performance (Tables 8.3, 8.4, and 8.5). The lowest level of performance is a building with envelope conditions that would not pass code in many countries but that are still allowed in many countries in the world without energy codes. The effect of implementing two passive design strategies (direct gain in the cold climate and night ventilation in the hot and dry) is also calculated. The tightness of the envelope is constant at a maximum of one air change per hour, which is not difficult to achieve. This value was not modified so as to not introduce another variable in the design, even though the more insulated buildings built with structural insulated panels (SIPs) would probably be much tighter, improving its performance.

GHG emissions from heating (Table 8.6) clearly demonstrate the effect of a better envelope on performance. In the cold climate, the house with the low-performance furnace (72% annual fuel utilization efficiency [AFUE]), poor insulation, and single-glazed windows, has emissions of 13,466 kg CO_2e. When the furnace is updated to 97% AFUE and the house is super-insulated, the emissions are reduced to 1,091 kg CO_2e, which is only 8.1% of the previous amount. If the windows are relocated to the south side and reduced on the east, west, and north sides; if an overhang is added on the south side; and if the floors are changed to concrete slabs, the emissions are further reduced to only 291 kg CO_2e (Figure 8.18), which is only 2.2% of the initial emissions in the low-performance house. Implementing passive strategies improves performance at all levels. Adding these three additional "passive solar" changes (windows to the south, overhangs, and concrete slabs) reduces the emissions by about 9% in the low-quality envelope, 37% in the code-quality envelope, and 73% in the high-quality envelope.

Emissions from cooling are also affected by the quality of the envelope and the efficiency of the heating, ventilating, and air-conditioning (HVAC) system. In the hot and dry climate, the emissions with the low-performance air-conditioning system (8.9 SEER) in the poorly insulated house with single-glazed windows are 9,523 kg CO_2e. When the air conditioning is updated to 18 SEER and the house is

TABLE 8.3
Envelope Characteristics

Envelope Components (W/m² K)	Wall	Roof	Floor
Low insulation	1.46	1.53	0.53
Medium insulation as defined by the California Title 24 Energy Code	0.27	0.15	0.29
Super-insulated Passivhaus Standard	0.15	0.13	0.15

TABLE 8.4
Window Characteristics

	SHGC	Transmissivity	U (W/m² K)
Single-glazed	0.75	0.77	7.2
Double-glazed	0.55	0.55	3.12
Triple-glazed	0.5	0.51	0.79

TABLE 8.5
Mechanical Systems

	Furnace (AFUE)	Air Conditioning (SEER)
Low-efficiency	72	8.9
Medium-efficiency	78	13
High-efficiency	97	18

super-insulated (see Table 8.5), the emissions are reduced to 1,994 kg CO_2e, which is only 20.9% of the previous amount (Table 8.7). If night ventilation was added to the best envelope with the highest-efficiency air-conditioning system, the emissions would be further reduced to 732 kg CO_2e, which is only 7.7% of the original value (Figure 8.19). Night ventilation, shading of the south windows, and thermal mass reduced emissions in all building qualities in the hot and dry climate. Emissions were reduced by about 26% in the poorly insulated building, 55% in the code building, and 63% in the super-insulated building.

Just applying these design strategies (better envelope, more efficient mechanical and passive strategies) can reduce emissions from 13,466 kg CO_2e in the uninsulated building with bad windows and inefficient mechanical system to 291 kg CO_2e with the super-insulated building in the cold climate and from 9,523 kg CO_2e to 732 kg CO_2e in the hot and dry climate, clearly demonstrating the power of energy-efficient and climate-responsive design.

TABLE 8.6
GHG Heating Emissions (kg CO_2e)

Heating

Climate	Cold			Temperature			Hot and Dry			Hot and Humid		
Furnace Efficiency AFUE	Low	Medium	High	Low	Medium	High	Low	Medium	High	Low	Medium	High
No insulation and single-glazed	13466	12429	9991	2596	2395	1928	1710	1575	1265	261	239	190
Code insulation and double-glazed	3817	3524	2829	407	369	299	358	331	266	38	33	27
Super-insulated and triple-glazed	1477	1363	1091	16	16	16	33	33	27	5	5	5

TABLE 8.7
GHG Emissions from Cooling (kg CO_2e)

Climate	Cold			Temperature			Hot & Dry			Hot & Humid		
Cooling SEER	Low	Medium	High	Low	Medium	High	Low	Medium	High	Low	Medium	High
No insulation and single-glazed	1893	1296	936	1784	1221	882	9523	6519	4708	6597	4517	3262
Code insulation and double-glazed	1127	771	557	1238	848	612	5100	3491	2522	4994	2741	1980
Super-insulated and triple-glazed	1259	862	622	1605	1091	787	4032	2748	1994	757	757	757

Passive Heating

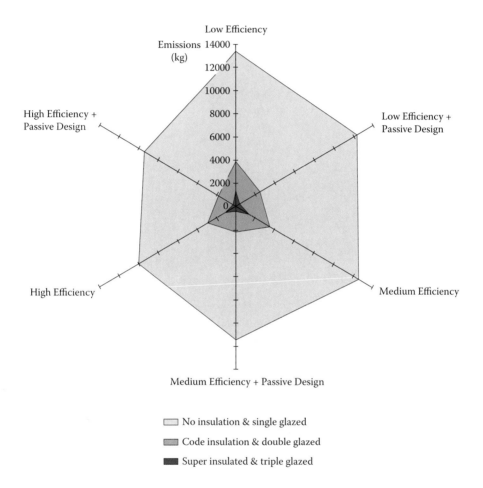

FIGURE 8.18 Emissions in cold climate with different strategies including passive strategies (direct solar gain).

288 Carbon-Neutral Architectural Design

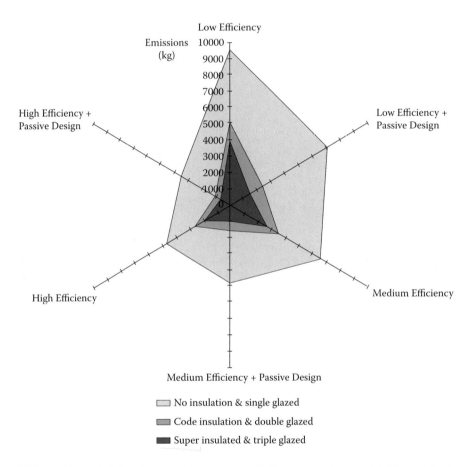

FIGURE 8.19 Emissions in hot and dry climates including passive strategies (night ventilation).

References

CHAPTER 1

Architecture 2030. (2009). http://www.architecture2030.org/current_situation/building_sector.html

B. Bordass, from Solar Cities: the Fundamental Documents. S Roaf and R Gupta. With contributions from Dr Bill Bordass, Chiel Boonstra, Catherine Bottrill and Robert Cohen. An outcome of the meetings of the Carbon Counting Working Group, London UK, 2006.

Boake, T., Carbon Neutral Design Project, contribution to the carbon neutral design project. Retrieved August 2010 from http://www.architecture.uwaterloo.ca/faculty_projects/terri/carbon-aia/tools.html

Brown, M. A., Stovall, T. K., and Hughes, P. J. (2007). Buildings, in *Tackling Climate Change in the U.S.: Potential Carbon Emissions Reductions from Energy Efficiency and Renewable Energy by 2030* (pp. 51–68). C. F. Kutscher (Ed.). Washington, D.C.: American Solar Energy Society.

Bryan H. and Trusty W. (2008). Developing an Operational and Material CO2 Calculation Protocol for Buildings. Paper presented at the Sustainable Building 2008 Conference, Melbourne, Australia, September 21–25.

Carbon Trust. (2007). Carbon Footprint Measurement Methodology, Version 1.3. London: Author.

Department for Environment Food and Rural Affairs. (DEFRA). (2005, July). Fuel Emission Factors. http://www.defra.gov.uk/environment/business/reporting/conversion-factors.htm

Department for Environment Food and Rural Affairs. (DEFRA). (2010, June 6). 2010 Guidelines to Defra/DECC's GHG Conversion Factors for Company Reporting. Retrieved September 2010 from http://archive.defra.gov.uk/environment/business/reporting/pdf/101006-guidelines-ghg-conversion-factors.pdf

Intergovernmental Panel on Climate Change. (IPCC). (2007). *Climate Change 2007: Synthesis Report. Contribution of Working Groups I, II and III to the Fourth Assessment Report of the Intergovernmental Panel on Climate Change*. (Core Writing Team, R. K. Pachauri, and A. Reisinger, Eds.). Geneva, Switzerland: Author.

IPCC (2007). Summary for Policymakers, in *Climate Change 2007: Impacts, Adaptation and Vulnerability. Contribution of Working Group II to the Fourth Assessment Report of the Intergovernmental Panel on Climate Change* (p. 17). Cambridge, UK: Cambridge University Press.

Krebs, M. (2007, February). Water Related Energy Use in California. Assembly Committee on Water, Parks and Wildfire. Retrieved July 2009 from http://www.energy.ca.gov/2007publications/CEC-999-2007-008/CEC-999-2007-008.PDF

La Roche, P. (2010). Carbon Counting in Architecture: A Comparison of Carbon Estimating Tools. *Informes de la Construccion* 62(517), Madrid, Spain.

La Roche, P. and Campanella, C. (April 24, 2009). Carbon Counting in Architecture: A Comparison of Several Tools. Paper presented at the American Solar Energy National Conference. Buffalo, NY.

La Roche, P. and Campanella, C. The Carbon Neutral Design Project, Carbon Calculation Tools, Survey of Carbon Calculators Retrieved September 2009 from http://www.architecture.uwaterloo.ca/faculty_projects/terri/carbon-aia/tools3.html

Malin, N. (2008, July 1). Counting Carbon: Understanding Carbon Footprints of Buildings. *Environmental Building News.*

Mazria, E. and Kershner, K. (2008, June 20). Meeting the 2030 Challenge through Building Codes. http://www.architecture2030.org/pdfs/2030Challenge_Codes_WP.pdf

Metz, B., Davidson, O., Bosch, P. R., Dave, R., and Meyer, L. (Eds.). (2007). *Climate Change 2007: Working Group III: Mitigation of Climate Change.* Cambridge, UK: Cambridge University Press.

Milne, M. (July 13, 2007). A Design Tool for Meeting the 2030 Challenge: Measuring CO2, Passive Performance, and Site Use Intensity. Paper presented at the American Solar Energy Association Conference, Cleveland, OH.

Mongameli Mehlwana, A., Mirasgedis, S., Novikova, A., Rilling, J., and Yoshino, H. (2007). Residential and Commercial Buildings. In: *Climate Change 2007: Impacts, Adaptation and Vulnerability. Contribution of Working Group II to the Fourth Assessment Report of the Intergovernmental Panel on Climate Change,* M. L. Parry, O. Canziani, J. P. Palutikof, P. J. van der Linden, and C. E. Hanson (Eds.). Cambridge, UK: Cambridge University Press.

Padgett, P., Stenemann, A., Clarke, J., and Vanderbergh, M. (2008). A Comparison of Carbon Calculators. *Environmental Impact Assessment Review* 28, 106–115.

Parry, M. L., Canziani, O. F., Palutikof, J. P., van der Linden, P. J., and Hanson, C. E. (Eds.). (2007). *Contribution of Working Group II to the Fourth Assessment Report of the Intergovernmental Panel on Climate Change.* Cambridge, UK: Cambridge University Press.

Rees, W. E. (1992, October). Ecological Footprints and Appropriated Carrying Capacity: What Urban Economics Leaves Out. *Environment and Urbanisation* 4, 121–130.

U.S. Environmental Protection Agency. (EPA). (2011a). eGRID Version 2.1. Year 2007 Summary Tables.

http://www.epa.gov/cleanenergy/documents/egridzips/eGRID2010V1_1_year07_SummaryTables.pdf

U.S. Environmental Protection Agency. (EPA). (2011c). Household Emissions Calculator. http://www.epa.gov/climatechange/emissions/ind_calculator.html

U.S. Environmental Protection Agency. (EPA). (2011b). Greenhouse Gas Emissions. http://www.epa.gov/climatechange/emissions/index.html#ggo

U.S. Energy Information Administration, Vehicle-Miles Traveled in the U.S. http://www.eia.doe.gov/emeu/rtecs/chapter3.html (accessed July 2009)

U.S. Environmental Protection Agency. (EPA). (2011d). Transportation and Climate. Retrieved September 2010 from http://www.epa.gov/OMS/climate/index.htm

United Nations Environment Programme (UNEP) Sustainable Buildings and Climate Initiative (SBCI). (2010). Common Carbon Metric for Measuring Energy Use and Reporting Greenhouse Gas Emissions from Building Operations. Retrieved May 26, 2011 from http://www.unep.org/sbci/pdfs/UNEPSBCICarbonMetric.pdf

Wackernagel, M., Monfreda, C., Moran, D., Wermer, P., Goldfinger, S., Deumling, D., et al. (2005). National Footprint and Biocapacity Accounts 2005: The Underlying Calculation Method. Oakland, CA: Global Footprint Network.

Yohe, G. W., Lasco, R. D., Ahmad, Q. K., Arnell, N. W., Cohen, S. J., Hope, C., et al. (2007). Perspectives on Climate Change and Sustainability. In: *Climate Change 2007: Impacts, Adaptation and Vulnerability. Contribution of Working Group II to the Fourth Assessment Report of the Intergovernmental Panel on Climate Change,* M. L. Parry, O. Canziani, J. P. Palutikof, P. J. van der Linden, and C. E. Hanson (Eds.). Cambridge, UK: Cambridge University Press, 811–841.

CHAPTER 2

An Assessment of the Intergovernmental Panel on Climate Change, Climate Change 2007: Synthesis Report, Valencia, Spain, November 12–17, 2007, Working Group contributions to the IV Assessment Report.

Architecture 2030. (2009). Retrieved August 2010 from http://www.architecture2030.org/current_situation/building_sector.html

http://www.athenasmi.org/tools/ecoCalculator/index.html

Boake, T. Carbon Neutral Design Project. Retrieved July 2009 from http://www.architecture.uwaterloo.ca/faculty_projects/terri/carbon-aia/tools.html

Boake, T. M. (2008). The Leap to Zero Carbon, Defining the First Steps to Carbon Neutral Design. Paper presented at SBSE Retreat, UK.

California Assembly Committee on Water, Parks and Wildlife. (2009). Study. http://www.energy.ca.gov/2007publications/CEC-999-2007-008/CEC-999-2007-008.PDF

Carbon Neutral Design Project. Retrieved July 2010 from aia.org/carbonneutraldesignproject

Carbon Trust. (2007). Carbon Footprint Measurement Methodology, Version 1.1. The Carbon Trust, London, UK. http://www.carbontrust.co.uk. Cited in "A Definition of 'Carbon Footprint'" 07-01 Wiedmann, T., Minx J., ISAUK Research (2007).

Climate Trust. CarbonCounter. Retrieved June 2009 from http://www.carboncounter.org/

eGRID2006 Version 2.1. (2007, April). Summary Tables. http://www.epa.gov/cleanenergy/energy-resources/egrid/index.html

Emission Facts: Average Carbon Dioxide Emissions Resulting from Gasoline and Diesel Fuel. epa.gov/OMS/climate/420f05001.htm

Epp, J., La Roche P., Fox, M. A., et al (2008). The PRIME Evaluation System: A Student Developed Eco Analysis Tool. Paper presented at the American Solar Energy National Conference, San Diego, CA.

Givoni, B. (1998). *Climate Considerations in Building and Urban Design.* John Wiley and Sons, NY, NY.

Hansanuwat, R. Lyles, M., West, M., and La Roche, P. (2007). A Low Tech–Low Cost Sustainable House for Tijuana, Mexico. Paper presented at the American Solar Energy National Conference, Cleveland, OH.

Jeerage, A., La Roche P., and Spiegelhalter, T. (2008). Low Cost Window Systems for the Tijuana House Prototype, Paper presented at the American Solar Energy National Conference, San Diego, CA.

La Roche, P. (1995). Herramienta Automatizada para el Diseño Bioclimatico de Edificaciones: ASICLIMA. *Tecnologia y Construccion* 11, 19–29.

La Roche, P. (2007a). Design for a Changing Climate: Teaching Ecological Architecture at Cal Poly Pomona University. Paper presented at the Teaching in Architecture International Conference, Krems.

La Roche, P. (2007b). Teaching Architecture in a Changing Climate. Paper presented at the Teaching in Architecture Conference, Krems, Austria.

La Roche, P. (2008). Teaching Climate Responsive Design to Beginning Students. Paper presented at the Teaching in Architecture International Conference, Oxford, UK.

La Roche, P. and Campanella, C. (2009). Carbon Counting in Architecture: A Comparison of Several Tools. Paper presented at the American Solar Energy National Conference, Buffalo, NY.

La Roche, P. and Liggett, R. (2001). Very Simple Design Tools: A Web Based Assistant for the Design of Climate Responsive Buildings. *Architectural Science Review* 44, 437–448.

Lyle, J. T. (1994). Regenerative Design for Sustainable Development, Wiley and Sons, NY, NY.

Mazria, E. (2008). Meeting the 2030 Challenge through Building Codes. Edward Mazria, Executive Director, Kristina Kershner, Director 2030, Inc./Architecture 2030, June 20, 2008 at http://www.architecture2030.org/pdfs/2030Challenge_Codes_WP.pdf

Milne, M. (2007). A Design Tool for Meeting the 2030 Challenge: Measuring CO2, Passive Performance, and Site Use Intensity. Paper presented at the American Solar Energy Association Conference, Cleveland, OH.

Milne, M., Bapat, A., Gomez, C., Han, Z., Kobayashi, K., Leeped, D., et al. (2000). Tools for Teaching Climate Responsive Design. Paper presented at the 3rd International Conference for Teachers of Architecture, Oxford, UK.

Milne, M., Liggett, R., and Alshaali, R. (2007). Climate Consultant 3.0: A Tool for Visualizing Building Energy Implications of Climates. Paper presented at the Annual Conference of the American Solar Energy Society, Cleveland, OH.

Milne, M., et al. (2001, June). A Drag and Drop Energy Design Tool. Paper presented at the Proceedings of the American Solar Energy Conference, Washington, DC.

Padgett, P., Stenemann, A., Clarke, J., and Vanderbergh, M. (2008). A Comparison of Carbon Calculators. *Environmental Impact Assessment Review* 28, 106–115.

Personal Emissions Calculator. (2009). EPA.

La Roche, P., Quirós, C., Bravo, G., Machado, M., and Gonzalez, G. (2001). *Plea Note 6, Keeping Cool: Principles to Avoid Overheating in Buildings*. Kangaroo Valley, Australia: Passive Low Energy Architecture Association & Research Consulting and Communications.

Rittel, H. (1970). Some Principles for the Design of an Educational System for Design. *Design Methods and Theories* 20, 359–375.

Roaf, S. and Gupta, R. (2006). *Solar Cities: The Fundamental Documents*.

Rowe, P. (1987). *Design Thinking*. Cambridge, MA: MIT Press.

Szokolay, S. (1996). *Solar Geometry*. University of Queensland, Australia. PLEA Notes 1, Design Tools and Techniques.

United Nations Environment Programme (UNEP) Sustainable Buildings and Climate Initiative (SBCI). (2009). Common Carbon Metric for Measuring Energy Use & Reporting Greenhouse Gas Emissions from Building Operations, Sustainable Buildings & Climate Initiative.

U.K. Department for Environment Food and Rural Affairs (DEFRA). (2005, July). Fuel Emission Factors.

U.K. Department for Environment Food and Rural Affairs. (DEFRA). (2010). Guidelines to DEFRA/DECC's GHG Conversion Factors for Company Reporting. Retrieved September 2010 from http://www.defra.gov.uk/environment/business/reporting/pdf/100805-guidelines-ghg-conversion-factors.pdf

U.S. Census Tract Information. Retrieved September 2007 from http://www.census.gov./geo/www/gazetteer/places2k.html

U.S. Energy Information Agency. (2009). Architecture 2030. http://www.architecture2030.com/current_situation/building_sector.html

U.S. Environmental Protection Agency (EPA). Personal Emissions Calculator. Retrieved September 2010 from http://www.epa.gov/climatechange/emissions/ind_calculator.html

U.S. Environmental Protection Agency (EPA). Waste Reduction Model (WARM). Retrieved September 2010 from http://www.epa.gov/climatechange/wycd/waste/calculators/Warm_home.html

U.S. Environmental Protection Agency (EPA). Climate Change and Municipal Solid Waste (MSW). Retrieved July 2009 from http://www.epa.gov/osw/conserve/tools/payt/tools/factfin.htm

West, M. and La Roche, P. (2008). Developing a Low Cost, Sustainable Housing Prototype Using Recycled Waste Materials in Tijuana, Mexico. Paper presented at the Passive Low Energy Architecture Conference PLEA, Dublin, Ireland.

Yezell, E., Felton, L., La Roche, P., and Fox, M. (2007). Greenkit: A Modular Variable Application Cooling System. Paper presented at the Annual Conference of the American Solar Energy Society, Cleveland, OH.

CHAPTER 3

American Society of Heating, Refrigerating, and Air-Conditioning Engineers. (ASHRAE). (2004). *ASHRAE Standard 55-2004 for Thermal Environmental Conditions for Human Occupancy.* Author.

American Society of Heating, Refrigerating, and Air-Conditioning Engineers. (ASHRAE). (2005). *Handbook of Fundamentals.* Author.

Auliciems, A. (1981). Towards a Psycho-Physiological Model of Thermal Perception. *International Journal of Biometeorology* 25, 109–122.

Auliciems, A. and Szokolay, S. (1997). PLEA Notes 3: Thermal Comfort.

Brager, G. and de Dear, R. J. (2000, October). A Standard for Natural Ventilation. *ASHRAE Journal.*

De Dear, R. J. and Brager, G. S. (1998). Developing an Adaptive Model of Thermal Comfort and Preference. *ASHRAE Transactions* 104(1a), 145–167.

De Dear, R. J. and Brager, G. S. Thermal Comfort in Naturally Ventilated Buildings: Revisions to ASHRAE Standard 55, *Energy and Buildings* 34, 549–561.

Fanger, P. O. (1967). Calculation of thermal comfort: Introduction of a basic comfort equation. *ASHRAE Transactions*, 73(2), III4.1-III4.20.

Fanger, P. O. (1994). How to Apply Models Predicting Thermal Sensation and Discomfort in Practice. In *Thermal Comfort: Past, Present and Future,* N. A. Oseland and M. A. Humphreys (Eds.). BRE Report.

Fanger, P. O. and Toftum, J. (2002). Extension of the PMV Model to Non-Air Conditioned Buildings in Warm Climates, *Energy and Buildings* 34(6), 533–536.

Fitch, J. M. (1947). American Building: The Environmental Forces That Shape It.

Griffiths, I. (1990). Thermal Comfort Studies in Buildings with Passive Solar Features: Field Studies, Report to Commission of the European Community, ENS35 909, UK.

Humphreys, M. A. (1975). Field Studies of Thermal Comfort Compared and Applied. *Journal of the Institute of Heating and Ventilation Engineering* 44, 5–27.

Humphreys, M. A. (1975). Field Studies of Thermal Comfort Compared and Applied. In: Department of the Environment: Building Research Establishment, CP 76/75.

Kwok, A. and Rajkovich, N. (2010). Addressing Climate Change in Comfort Standards. *Building and Environment* 45, 18–22.

National Oceanic and Atmospheric Administration. (NOAA). (2010). Mauna Loa CO_2 Annual Mean Data. ftp://ftp.cmdl.noaa.gov/ccg/co2/trends/co2_mm_mlo.txt

Nicol, F. and Humphreys, M. A. (2010). Derivation of the Adaptive Equations for Thermal Comfort in Free Running Buildings in European Standard EN15251, *Building and Environment* 45.

Nicol, J. F. and Humphreys, M. A. (2010). Derivation of the Equations for Comfort in Free-Running Buildings in CEN Standard EN15251. *Special Issue Section: International Symposium on the Interaction Human and Building Environment, Buildings and Environment* 45(1), 11–17.

Nicol, J. F. and Roaf, S. (1996). Pioneering New Indoor Temperature Standards: The Pakistan Project. *Energy and Buildings* 23, 169–174.

Rohles, F. (1973). The Revised Modal Comfort Envelope. *ASHRAE Transactions* 79(2), 52.

Rohles, F. and Nevins, R. (1971). The Nature of Thermal Comfort for Sedentary Man. *ASHRAE Transactions* 78(1), 131.

CHAPTER 4

Achard, P. and Gicquel, R. (Eds.). (1986). *European Passive Solar Handbook. Basic Principles and Concepts for Passive Solar Architecture*. Brussels: Commission of the European Communities, Directorate General XII for Science, Research and Development.

Anderson, B. (1981). *Passive Solar Design Handbook*, Volume 1. U.S. Department of Energy.

ASHRAE Standard 90.1 (2007). Energy Standard for Buildings Except Low-Rise Residential Buildings. American Society of Heating Refrigerating and Air Conditioning Engineers, Atlanta, GA.

Balcomb, D. (1992). *Passive Solar Buildings*. Cambridge, MA: MIT Press.

Bordass, B. (2006). Solar Cities: The Fundamental Documents. S. Roaf and R. Gupta (Eds.). An outcome of the meetings of the Carbon Counting Working Group, London UK.

Bryan, H. and Trusty, W. (2008). Developing an Operational and Material CO2 Calculation Protocol for Buildings. Paper presented at the SB08, Melbourne, Australia.

Building Energy Codes Resource Center. (2010). Energy Code Climate Zones, Article 1420. http://resourcecenter.pnl.gov/cocoon/morf/ResourceCenter/article/1420

California Commission on Energy Waters and Parks. http://www.energy.ca.gov/2007publications/CEC-999-2007-008/CEC-999-2007-008.PDF

Docherty M. and Szokolay S. (1999). *Climate Analysis, PLEA Notes, Design Tools and Techniques: Note 5*. Queensland, Australia: Department of Architecture, University of Queensland.

eGRID2006 Version 2.1 (2007, April). Summary Tables.

Evans, M. (2007). The Comfort Triangles: A New Tool for Bioclimatic Design. PhD thesis, Delf Technical University.

Gasparini, G. (1993). *Venezuelan Houses*. Caracas, Venezuela: Ernesto Armitano Editores.

Givoni, B. (1976). *Man Climate and Architecture*, 2d ed. London: Applied Science Publishers.

Givoni, B. (1992). Comfort, Climate Analysis and Building Design Guidelines. *Energy and Buildings* 18(1), 11–23.

Givoni, B. (1995). *Passive and Low Energy Cooling of Buildings*. New York: Van Nostrand Reinhold.

Givoni, B. (1998). *Climate Considerations in Building and Urban Design*. New York: Van Nostrand Reinhold.

Givoni, B., Gulish, M., Gomez, C., and Gomez, A. (1996). Radiant Cooling by Metal Roofs in Developing Countries. In: *21st National Passive Solar Conference*, Asheville, NC, 83–87.

Gonzalez, E., Hinz, E., Oteiza, P., and Quiros, C. (1986). *Proyecto Clima y Arquitectura*. Ediciones G Gili Mexico.

Koenigsberger, O. H., Mahoney, C., and Evans, M. (1971). *Climate and House Design*. United Nations: NY. Part 1 (pp. 3–8), Part 2 (pp. 11–93) prepared by the Center for Housing, Building, and Planning of the United Nations Dept. of Economic and Social Affairs

Köppen, W. (1918). Klassifikation der Kliamte nach Temperatur, Nierderschlag und Jahreslauf. *Petermanns Mitt*, 64, 193–203.

La Roche, P. (1992). Design recommendations for housing in warm humid climates. In: *Proc. of Energy Environment and Technological Innovation, 2nd International Congress*, Rome, Italy, 1, pp. 319–324.

La Roche, P. (2002). The case study house program in Los Angeles: A case for sustainability. In: *Proc. of PLEA 2002 Conference (Passive and Low Energy Architecture)*, Toulouse, France.

La Roche, P. (2010a). *Carbon Counting in Architecture: A Comparison of Carbon Estimating Tools*. Madrid, Spain: Informes de la Construccion, Informes de la Construccion.

La Roche, P. (2010b, May). Design Strategies for Low Emissions. *Solar Today* 24(4).

La Roche, P. (2010, May). Design strategies for low emissions. *Solar Today*, 24(4).

La Roche, P. and Almodóvar, J. (2009). El Case Study House Program en Los Angeles: Un ejemplo de Sostenibilidad. "Revista de Historia y Teoría de la Arquitectura" n° 8–9. Funded by the Consejería de Cultura de la Junta de Andalucía. ISSN: 1576–5628.

Lewis, O J. (Ed.). (2001). *A Green Vitruvius: Principles and Practice of Sustainable Architectural Design.* Thermie Program of the Ec Dgxvii.

Machado, M., La Roche, P., Mustieles, F., and De Oteiza, I., (2000). The fourth house: The design of a bioclimatic house in Venezuela, *Building Research Information,* 28(3), 196–211.

McCoy, E. (1989). in Smith, Elizabeth. Blueprints for Modern Living. History and Legacy of the Case Study Houses. Edited by. 1989. The Museum of Contemporary Art. Los Angeles.

Milne, M. (2007). A Design Tool for Meeting the 2030 Challenge: Measuring CO2, Passive Performance, and Site Use Intensity. Paper presented at the American Solar Energy Association Conference, Cleveland, OH.

Milne, M. and Givoni, B. (1979). Architectural Design Based on Climate. In: *Energy Conservation through Building Design,* D. Watson (Ed.). New York: McGraw Hill, 96–113.

Moffat, A.S. and Schiler, M. *Landscape Design That Saves Energy,* New York: William Morrow and Co., 1981.

Olgyay, V. (1963). *Design with Climate, A Bioclimatic Approach to Architectural Regionalism.* Princeton, NJ: Princeton University Press.

Oliver, J. (1991). The history, status and future of climatic classification. *Physical Geography* 12(3), 231–251.

Rapoport, A. (1969), *House Form and Culture.* Upper Saddle River, NJ: Foundations of Cultural Geography Series, Prentice Hall.

Reynolds, J. (1992). *Courtyards: Aesthetic Social and Thermal Delight.* New York: John Wiley and Sons.

U.K. Department for Environment Food and Rural Affairs (DEFRA). (2005, July). Fuel Emission Factors.

U.S. Energy Information Administration. (EIA). Vehicle Miles Travelled. http://www.eia.doe.gov/emeu/rtecs/chapter3.html (accessed 2009).

U.S. Environmental Protection Agency. (EPA). Emission Facts: Average Carbon Dioxide Emissions Resulting from Gasoline and Diesel Fuel.

U.S. Environmental Protection Agency. (EPA). Household Emissions Calculator. http://www.epa.gov/climatechange/emissions/ind_calculator.html (accessed 2009).

Watson, D. and Labs, K. (1983). *Climatic Building Design: Energy-Efficient Building Principles and Practices.* New York: McGraw-Hill.

CHAPTER 5

Lewis, O., *A Green Vitruvius: Principles and Practice of Sustainable Architectural Design,* 2nd ed., 2001. James and James, London.

CHAPTER 6

Achard, P. and Qicquel, R. (1986). *European Passive Solar Handbook. Basic Principles and Concepts for Passive Solar Architecture. Premilinary Edition.* Brussels: Editado por Commission of the European Communities. Directorate - General XII for Science, Research and Development.

Bansal, N., Hauser, G., and Minke, G. (1994). *Passive Building Design: A Handbook of Natural Climatic Control.* Amsterdam: Elsevier Science.

Cook, J. (1986). *Passive Cooling.* Cambridge, MA: The MIT Press.

Da Veiga, J. and La Roche, P. (2003). A Computer Tool for the Analysis of Direct Solar Radiation in Complex Architectural Envelopes: EvSurf. 9th Conference on Building Science and Technology, February 27, 28, 2003, Vancouver, BC, Canada.

Fernandes, J. and La Roche, P. (2002). A Computer Tool for the Design and Manufacturing of Complex Architectural Envelopes with Solar Performance. Paper presented at the Sixth Conference of the Iberoamerican Society of Digital Graphics, Caracas, Venezuela.

Ghiabaklou, Z. (2010). Natural ventilation as a design strategy for energy saving. *World Academy of Science, Engineering and Technology*, 71.

Givoni, B. (1976). *Man, Climate and Architecture*, 2nd ed. London, UK: Applied Science Publishers.

Givoni, B. (1994). *Passive and Low Energy Cooling of Buildings*. New York: Van Nostrand Reinhold.

González, E. (1997). Étude de matériaux et de techniques de refroidissement passif pour la conception architecturale bioclimatique en climat chaud et humide. Thèse de doctorat en Energétique de l'Ecole des Mines de Paris, France.

Izard, J-L. (1993). *Architectures d'été. Construire pour le confort déte*. La Calade, France: Edisud.

Koenigsberger, O. H., Ingersoll, T. G., Mayhew, A., and Szokolay, S. V. (1977). *Viviendas y edificios en zonas cálidas y tropicales*. Madrid: Editorial Paraninfo.

La Roche, P. and Fernandes, J. (2003). A Computer Tool for the Analysis of Direct Solar Radiation in Complex Architectural Envelopes: EvSurf. Paper presented at the 9th conference on Building Science and Technology, February 27–28, Vancouver, BC.

La Roche, P. and Milne, M. (2004). Effects of Window Size and Mass on Thermal Comfort using an Intelligent Ventilation Controller. *Solar Energy* 77, 421–434.

La Roche, P., Quirós, C., Bravo, G., Machado, M., and Gonzalez, G. (2001). *Plea Note 6, Keeping Cool: Principles to Avoid Overheating in Buildings*. Kangaroo Valley, Australia: Passive Low Energy Architecture Association & Research Consulting and Communications.

Lavigne, P., Brejon, P., and Fernandez, P. (1994). *Architecture Climatique. Une contribution au developpement durable. Tome 1: Bases Physiques*. La Calade, France: Edisud.

Lewis, O J. (Ed.). (2001). *A Green Vitruvius: Principles and Practice of Sustainable Architectural Design*. Thermie Program of the Ec Dgxvii.

McPherson, E. G. (1984). *Energy Conserving Site Design*. American Society of Landscape Architects.

Olgyay, A. and Olgyay, V. (1957). Solar Control and Shading Devices. Princeton, NJ: Princeton University Press.

Olgyay, V. (1968). *Clima y Arquitectura en Colombia*. Cali: Universidad del Valle.

Sobin, H. J. (1981). Window design for passive ventilative cooling: An experiment model scale study. In *Proc. International Passive/Hybrid Cooling Conference*, Miami Beach, FL.

Watson, D. and Labs, K. (1983). *Climatic Design. Energy Efficient Building Principles and Practice*, 2nd Edition. New York: McGraw-Hill.

Zold, A. and Szokolay, S. (1997). Thermal Insulation. Plea Note 2. 66 p.

CHAPTER 7

Beauchamp, H. (2010). Buildings that Heat and Cool Themselves: Passive Double Play in the First Two Years of Operation. Paper presented at the ASES Conference.

Blondeau, P., Sperandio, M., and Allard, F. (2002). Night Ventilation for Building Cooling in Summer. *Solar Energy* 61(5), 327–335.

Cook, J. (1986), *Passive Cooling*. Cambridge, MA: The MIT Press.

Cunningham, W.A. and Thompson, T. L. (1986). Passive cooling with natural draft cooling towers in combination with solar chimneys. *Proceedings Passive Low Energy Architecture* (PLE) Pecs, Hungary.

Del Barrio, E. (1998). Analysis of the Green Roofs Cooling Potential in Buildings. *Energy and Buildings* 27, 179–193.

References

Eftekhari, M. and Marjanovic L. (2003). Application of fuzzy control in naturally ventilated buildings for summer conditions. *Energy and Buildings*, 35, 645–655.

Eumorfopoulou, E. and Aravantinos, D. (1998). The Contribution of a Planted Roof to the Thermal Protection of Buildings in Greece. *Energy and Buildings* 27, 29–36.

Fernandez-Gonzalez, A. and Hossain, A. (2010). Cooling Performance and Energy Savings Produced by a Roofpond in the United States Southwest. American Solar Energy Society National Conference, Phoenix.

Geros, V., Santamouris, M., Tsangasoulis, A., Guarracino, G. (1999). Experimental Evaluation of Night Ventilation Phenomena, *Energy and Buildings* 29, 141–154.

Givoni, B. (1992). Comfort, climate analysis and building design guidelines. *Energy and Buildings*, 18(1), 11–23.

Givoni, B. (1994). *Passive and Low Energy Cooling of Buildings*. New York: Van Nostrand Reinhold.

Givoni B. (1998). *Climate Considerations in Building and Urban Design*. New York: Van Nostrand Reinhold.

Givoni B. (2011). Indoor temperature reduction by passive cooling systems. *Solar Energy*, 85, 1692–1726.

Givoni, B., Gulish, M., Gomez, C., and Gomez, A. (1996). Radiant cooling by metal roofs for developing countries. In: *Proceedings of the 21st National Passive Solar Conference*, Asheville, NC, April.

Givoni, B. and La Roche, P. (2001). Incidence of the Distribution of Mass in the Air Temperature of a Simple Roof Radiator. In: *PLEA 2001*, Florianopolis, Brazil, pp. 803–807.

Givoni, B. and La Roche, P. (2002). Modeling a Radiant Cooling Test Cell with Different Ua Values. In: *Proceedings of PLEA 2002 Conference (Passive and Low Energy Architecture)*, Toulouse, France.

Gonzalez, E., Machado, M. V., Rodriguez, L., Leon, G., Soto, M. P., and Almao, N. (2000). VBP-1: A sustainable urban house for low-income family in a tropical climate. PLEA 2000. Cambridge, UK.

Gronzik, W. T., Kwok, A. G., Stein, B., and Reynolds, J.S. (2010). *Mechanical and Electrical Equipment for Buildings*, 11th ed. Hoboken, NJ: John Wiley & Sons.

Hansanuwat, R., Lyles, M., West, M., and La Roche, P. (2007). A Low Tech—Low Cost Sustainable House for Tijuana, Mexico. Paper presented at the American Solar Energy National Conference, Cleveland, OH.

Kruger, E., Gonzalez, E., and Givoni, B. (2010). Effectiveness of indirect evaporative cooling and thermal mass in a hot arid climate. *Building and Environment*, 45, 1422–1433.

La Roche, P. (2002). Passive Cooling Systems for Sustainable Architecture in Warm Developing Countries. Paper

La Roche, P. (2005a). Effects of Thermal Mass, Smart Shading and Smart ventilation on indoor temperature. Paper presented at the 4th Latin American Conference on the Thermal Environment, May, Mexico City, Mexico.

La Roche, P. (2005b). Passive Cooling Systems for Sustainable Architecture in Developing Countries. Using the Gap to Bridge the Gap. Paper presented at the World Sustainable Building Conference. September 27–29, Tokyo, Japan.

La Roche, P. (2005c). Passive Smart Passive Cooling Systems: Tools for Sustainable Architecture. Paper presented at the International Conference on Modern Applied Technology and Management Science, North China University of Technology, August, Beijing.

La Roche, P. (2006a). Green Cooling: Combining Vegetated Roofs with Night Ventilation, *American Solar Energy*, Energy National Conference, Boulder, CO.

La Roche, P. (2006b). Green Cooling: Vegetated Roofs with Night Ventilation. Paper presented at the American Solar Society Conference, Denver, CO.

La Roche, P. (2006c). Smart Passive Systems: An Untapped Tool for Carbon Neutral Buildings, International Symposium on Sustainable Habitat Systems, Kyushu University, October 16–17, Fukuoka, Japan.

La Roche, P. (2009). Low Cost Green Roofs for Cooling: Experimental Series in a Hot and Dry Climate. Paper presented at the Passive Low Energy Conference, Quebec, Canada.

La Roche, P. (2010). Uninsulated Green Roofs: A Cooling Solution for Southern California. Paper presented at the Building Enclosure Sustainability Symposium, The Cal Poly Pomona Department of Architecture and Simpson Gumpertz & Heger.

La Roche, P. and Givoni, B. (2000). Indirect Evaporative Cooling with an Outdoor Pond. In: *Proceedings of the Solar 2000, the 25th Annual ASES Conference,* Madison, WI.

La Roche, P. and Givoni, B. (2000). Indirect Evaporative Cooling with an Outdoor Pond. In: *Proceedings of the Passive Low Energy Conference, Architecture City and the Environment,* Cambridge UK, pp. 310–311.

La Roche, P. and Givoni, B. (2001). Modeling Radiant Cooling Systems for Developing Countries. In: *Proceedings of ISES World Conference, Bringing Solar Down to Earth,* Adelaide, Australia.

La Roche, P. and Givoni, B. (2002). The Effect of Heat Gain on the Performance of a Radiant Cooling System. In: *Proceedings of the Passive and Low Energy Architecture Conference,* Toulouse, France.

La Roche, P. and Milne, M. (2001). Smart Controls for Whole House Fans. In: *Proc. of Cooling Frontiers, The Advanced Edge of Cooling Research and Applications in the Built Environment,* College of Architecture and Environmental Design, Arizona State University.

La Roche, P. and Milne, M. (2002). Effects of Thermal Parameters in the Performance of an Intelligent Controller for Ventilation. In: *Proceedings of the Annual Conference of the American Solar Energy Society,* Reno, NV.

La Roche, P. and Milne, M. (2003). Effects of Window Size and Mass on Thermal Comfort Using an Intelligent Ventilation Controller. American Solar Energy National Conference, Solar 2003, Austin, Texas, June 21–26.

La Roche, P. and Milne, M. (2004a). Automatic Sun Shades an Experimental Study. Paper presented at the American Solar Energy National Conference, Portland, OR, July 11–14.

La Roche, P. and Milne, M. (2004b). Effects of Window Size and Mass on Thermal Comfort using an Intelligent Ventilation Controller. *Solar Energy* 77, 421–434.

La Roche, P. and Milne, M. (2005). Effects of Smart Shading and Ventilation on Thermal Comfort Inside a Test Cell. Paper presented at the International Solar Energy Society and American Solar Energy National Conference, Orlando, Florida, August.

La Roche, P., Ramirez, I., Brown, K., Whitsett, K., Wehinger, K., Carranza, M., et al. (2006). A Very Low Cost Sustainable Housing Prototype for Tijuana, Mexico. Paper presented at the Passive Low Energy Architecture Conference, Geneva, Switzerland.

Machado, M. and La Roche, P. (1999). Construction Materials for Warm and Humid Climates (in Spanish: "Materiales de Construcción para Regiones de Clima Calido"). *Revista de Información Tecnológica,* 10(1), 243–250.

Marnich, R., Yamnitz, R., La Roche, P., and Carbonnier, E. (2010). Passive Cooling with Self-Shading Modular Roof Ponds as Heat Sink in Hot Arid Climates. Paper presented at the American Solar Energy National Conference, Phoenix, AZ.

Milne, M., Gomez, C., La Roche, P., and Morton, J., (2005), Comparing Three Passive Cooling Strategies in Sixteen Different Climate Zones. International Solar Energy Society and American Solar Energy National Conference, Orlando, Florida, August.

Niachou, A., Papakoustantinou, K., Santamouris, M., Tsangrassoulis, A., and Mihalakakou, G. (2001). Analysis of the Green Roof Thermal Properties and Investigation of Its Energy Performance. *Energy and Buildings* 33, 719–729.

References

Sharan, G. and Jadvav, R. (2003). Performance of Single Pass Earth-Tube Heat Exchanger: An Experimental Study. Retrieved from http://www.builditsolar.com/Projects/Cooling/Earth%20Tubes2003-01-07GirjaSharan.pdf

Shaviv E., Yezioro, A., and Capuleto, I. (2001). Thermal mass and night ventilation as a passive cooling design strategy. *Renewable Energy*, 24, 445–452.

Spindler H., Gliksman, L., and Norford L. (2002). The potential for natural and hybrid cooling strategies to reduce cooling energy consumption in the United States. Roomvent 2002 Copenhagen, Denmark.

Theodosiou, T. (2003). Summer Period Analysis of the Performance of a Planted Roof as a Passive Cooling Technique. *Energy and Buildings* 35, 909–917.

West, M. and La Roche, P. (2008). Developing a Low Cost, Sustainable Housing Prototype Using Recycled Waste Materials in Tijuana, Mexico. Paper presented at the Passive Low Energy Architecture Conference, Dublin, Ireland.

Yannas, S., Evyatar, E., and Molina, J. (2006). *Roof Cooling Techniques, A Design Handbook, Earthscan*. New York: Routledge.

CHAPTER 8

La Roche, P. (2009). Low Cost Green Roofs for Cooling: Experimental Series in a Hot and Dry Climate. Paper presented at the Passive Low Energy Conference, Quebec, Canada.

La Roche, P. (2010a). Carbon Counting: Simple Tools to Reduce Greenhouse Gas Emissions from Buildings. Building Enclosure Sustainability Symposium (BESS), The Cal Poly Pomona Department of Architecture and Simpson Gumpertz & Heger.

La Roche, P. (2010b). *Carbon Counting in Architecture: A comparison of Carbon Estimating Tools*. Madrid, Spain: Informes de la Construccion.

La Roche, P. and Campanella, C. (2009). Carbon Counting in Architecture: A Comparison of Several Tools. Paper presented at the American Solar Energy National Conference, Buffalo, NY.

Index

A

Absolute humidity, 79; *See also* Moisture content; Psychrometric charts
    climate data, 97
    defined, 75
    evaporative cooling, 242, 243
    heating/cooling air and, 77–78, 82
Absorbance, solar radiation, 191, 192
Absorption
    energy flow, sunshade, 198
    surface roughness and, 201
Absorptive materials, radiative heat exchange, 201–202, 203
Absorptivity
    colors, 193
    factors affecting solar radiation, 191
    insulating materials, 185
    solar radiation, 192
    and surface temperature, 197
Absorptivity factor, 193
Accessibility, design tools, 18
Acclimatization, physiological, 86
Active solar energy, 22
    design process roadmap, 21
    hot water, 54–55
    solar and site analysis, 39
Activity
    comfort triangle chart, 114, 115
    and heat balance, 85, 86, 87
    and thermal comfort, 93, 95
Act on CO_2 calculator, 12
Adaptive models of thermal comfort, 90, 92–95, 96
Adiabatic, defined, 78
Adiabatic heating/cooling, 83, 242, 243
Adjacent structures; *See* Massing of buildings; Surroundings, buildings and objects
Adobe, 121
Aerogel, silica, 206
Affordable housing, 34, 35, 36
Air
    conductivity, 184
    as heat sink, 221, 222; *See also* Evaporative cooling/heat loss; Ventilatory cooling
    infiltration; *See* Infiltration
    insulating materials, 184–185
Air conditioning, 230
Air exchange; *See also* Ventilation
    indoor air pollutants, 270
    open and closed loop systems, 266

Air exchange rate
    and heat transfer, 211
    night ventilation efficiency, 230
    smart ventilation model system, 233–234
Air flow/movement
    convective heat transfer, 101, 167, 210–220; *See also* Convective heat exchange
    effective airspeed, 224–226
    HMC Architects flex school building, 69
    natural ventilation, 51, 53, 54
    neighborhood design, 40
    and passive cooling, 268
    passive cooling
        nocturnal ventilation, 230–231
        smart ventilation model system, 232
    studio projects, advanced, 37
    studio projects, beginning year, 25, 26, 33
    and thermal comfort, 82, 91; *See also* Ventilatory cooling
    and evaporation, 84–85
    heat balance, human body, 102
    operative temperature, 86, 88
Air gaps/spaces
    conductive heat exchange, 185
    convective heat flows, 211
    and infiltration; *See* Infiltration
    thermal resistance of, 177
Air mass
    convective heat exchange, 211
    energy flow design, 166, 167
Air quality
    construction emissions reduction, 24
    indoor; *See* Indoor air quality
Air speed/velocity
    effective, 224–226
    and surface resistance, 185
    and thermal comfort, 90, 91
        comfort indices, 86
        convective heat exchange, 101
        evaporative cooling, 84–85, 102
    ventilation and passive cooling
        changing airflow direction and, 217
        comfort ventilation, 224–225, 226
        cooling towers, 245
        evaporative cooling, 215
        obstructions and, 219
        Venturi effect, 216
        window size and, 216
    wind; *See* Wind

Air temperature
  climate analysis
    building bioclimatic charts, 108–109
    Climate Consultant wind wheel, 113, 114
    climate types, 102–103
  comfort ventilation and, 226
  evaporative cooling, 242
  indoor/outdoor difference; *See* Temperature difference, indoor-outdoor
  radiative heat exchange, 259
  and thermal comfort, 82
    moisture content and, 78
    thermal neutrality temperature, 90, 92
Albedo
  ground, 207
  surrounding surface, 166
Altitude, and climate, 123, 124
Altitude, sun (solar geometry), 137, 139, 141, 159
Altitude angle, solar, 147, 152, 159
  shadow angle protractor, 149
  sun path diagrams, 142
  vertical shadow angle (VSA), 146, 147
Aluminum, 203
Aluminum window systems, 207
American Forests (analytical tool), 12, 13
American Society of Heating Refrigerating and Air Conditioning Engineers; *See* ASHRAE
American Solar Energy Society (ASES), 8
Analysis
  beginning year projects, 25–26
  modeling tools; *See* Digital tools/software
Andalucia, Spain, 121–123
Andes Mountains, 129
Angle of incidence, solar radiation, 138, 203
Angles, building components
  air inlets/outlets, 218–219
  slope of roof, 118, 220, 261
  solar gain factor, 202
Angles, shadow; *See* Shadow angles
Angles, solar
  bearing (azimuth), 139, 141, 200, 201
  hour, 137, 139, 140, 201
  modeling software, 200, 201
Annual cycles
  seasons; *See* Seasons
  solar geometry, 138, 139, 140
  sun position, 139, 140
Annual data, 112
  Climate Consultant, 113, 114
  climate data sets, 99
  climate effects simulations, 132
  timetable of bioclimatic needs, 113
Annual fuel use efficiency (AFUE), 284
Anthropogenic emissions; *See* Emissions
Apertures; *See* Openings, placement and size

Appliances
  energy flow design, 166, 167
  low-energy envelope design exercise, 44
  percentage of emissions in geographic subareas, 39
Architectural elements
  shadow angles, 149
  and solar heat gain coefficient, 204
Arid zone; *See* Hot dry climates
Art & Architecture Case Study Houses, 125–126, 128
ASHRAE
  climate data for design, 99
  thermal comfort
    definition of, 78
    prediction tools, 89; *See also* Psychrometric charts
ASHRAE standards
  climate classification systems, 103–104
  cluster analysis use by, 103
  digital climate analysis tools, 112
  thermal comfort, 90, 91, 92, 93, 94–95
Aspect ratio, courtyard, 122–123
Asphalt, 44
Athena: EcoCalculator for Assemblies, 12, 15
Atmosphere, as heat sink, 221, 222; *See also* Evaporative cooling/heat loss; Radiative heat exchange; Ventilatory cooling
Atmospheric pressure, 242
Atria, 49
Auliciems, A., 92
Australia, climate data sets, 99
Awnings, 145
Axis of earth rotation, 137, 138
Azimuth (bearing angle), 139, 141, 200, 201

B

Barajas, Gabriela, 36
Barriers
  air spaces, 185
  control of heat loss, 269
Baseline, emissions, 19
Batungbatal, Aireen, 52, 53
Beadwall, 210
Beam radiation, 166, 167, 202
Bearing angle (azimuth), 139, 141, 200, 201
Bearing angle (horizontal shadow angle), 146
Be Green Now, 12, 13
Behavioral adaptations, thermal comfort, 90, 92–95
Berkeley Institute of the Environment, 12
Best Foot Forward, 12
Bioclimatic chart, building, 106–111
Bioclimatic design, 221

Index

Biological carbon cycle, 1
Black bodies, 203
Black body radiation, 191, 193
Black surfaces, 202, 203
Blinds, 146, 210, 235, 274
Blood flow, 95
Blood vessels, temperature regulation, 80, 93
Body temperature; *See* Physiology/heat balance
Boilers, 9
Boiling, latent heat of evaporation, 172–173
Bonneville Environmental Foundation, 12, 13
Boundaries, operations emissions simulations, 130
BP Calculator, 12
Bravo, G., 165, 166, 198
Bricks, 180, 193
Bridges, thermal, 199–200
British thermal unit (Btu), 165, 171
Brown, Kyle, 34
Brunel, Jeremy, 31
Btu (British thermal unit), 165, 171
Budget, design process, 19
Build Carbon Neutral, 15
Build Carbon Neutral Construction Calculator, 12, 55
Building bioclimatic charts, 112, 113
Building properties/structure
    energy simulation tool exercises, 36, 37
    mass of; *See* Mass, structural/building
    passive cooling, 222
        heat transfer processes, 259
        nocturnal ventilation, 229
    radiative heat exchange, volume effects, 200
Buildings, adjacent; *See* Massing of buildings; Surroundings, buildings and objects
Built environment
    climate analysis, site-specific conditions, 99–100
    global warming effects, 6

C

CAD tools, 19
Calculators, carbon footprint, 10–11
CalEarth Ecodomes, 256, 258
California, climate effects on emissions simulations, 131
California Carbon Calculator, 12
California coast climate, 123, 124
California Energy Code comfort model, 112
California Energy Efficiency Standards for Residential and Nonresidential Buildings, 104, 105, 130, 132
California State Polytechnic University, Pomona green cooling model, 236–242
    HMC Architects building design, 74
California State Polytechnic University design program
    design process, 20, 21
    Pamo Valley project, 60–61, 62, 63, 64
    studio projects; *See* Studio projects
California State University classroom building design, 63, 65, 73
Camelback School, AZ, 249
Campanella, Charles, 10, 52
Canopy layer, green roof, 239, 241
Canziani, O.F., 3, 7
Capacitative insulators, 185
CarbonCounter.org, 12
Carbon counting tools, 10–14
    comparisons of emissions from natural gas and electricity, 12–14
    recommended, 14, 15
    selection of, 11, 12
    studio projects, advanced, 37
Carbon dioxide
    building emissions and environmental interactions, 8
    global warming potential, 2
    sources of emissions, 1
Carbon dioxide equivalents (CO_2e), 2, 11, 19
Carbon Footprint (design tool), 12
Carbon footprint calculators, 10–11, 12
Carbon Fund, 12, 13
Carbon intensity reduction, 20, 21, 22
Carbon intensity reduction, design process roadmap, 20, 21, 22
Carbon neutral architectural design (CNAD), 17–24
    design process, 17–25
        CAD tools, 19
        Cal Poly Pomona scheme, 20, 21
        classification of buildings, emitter/sequester, 19–20
        construction emissions, 23–24
        criteria, 19
        knowledge areas, 20, 21, 22, 23
        operations and energy emissions, 23
        phases of, 19
        traditional, 17, 18
        waste, emissions from, 24–25
        water, emissions from, 24
    integrated design process, 60–61, 62, 63, 64
    professional practice, 61, 63, 65–74
    studio projects, advanced, 28–29, 33, 34–60
        climate analysis, 38
        construction, 56
        daylighting, 49–50, 51, 52, 53

roof ponds, 249
    modular roof pond, 255–256, 257, 258
    smart roof pond with floating insulation, 254–255

Index

energy simulation tool exercises, 36–38
envelope, building, 41, 42, 43, 44
fenestration and shading, 44, 47–49, 50, 51
geographic distribution of emissions, 38–39, 40
Green kit, 29
housing prototypes, 34–35, 36
irrigation, 56
light shelves, 50–51
materials, low-carbon, 55, 56
modular design, 55
neighborhood design, 40–41
plants, use of, 56
radiation impact on surfaces, 42, 44, 45, 46
site analysis, 39, 40
solar gain, 53
solar hot water, 53–54
solar/photovoltaic systems, 55
ventilation, 51, 53, 54
waste management, 57, 60
water conservation, 56–57, 58, 59, 60
studio projects, beginning year, 25–28, 29–33
climate analysis, 28, 29
daylight analysis, 28, 31, 32, 33
envelope, building, 28
photovoltaic design, 28
Carbon sinks, 19
Case Study Houses, Los Angeles, 125–126, 128
Castro, David, 42
Ceilings, 49, 51, 131
Celestial hemisphere, 140–144
Celsius scale, 165
Cement, 188
Census tract data, 39
Center for Alternative Technologies, 12
Cervantes, Leslie, 44, 56, 57, 59
China, 29, 99
Chuck Wright (analysis tool), 12, 13
Churuata, 116
CIE standard overcast sky, 28
City, urban platform, 35, 36
City of Fair Oaks, emission area applications, 12
City of Tampa, emission area applications, 12
Classification of buildings (emitter/sequester), 19–20
Clay soil, 180
Clear Water (analysis tool), 12, 13
Clerestory windows, 42, 49, 68
Clima (climate analysis tool), 111–112
Climate, 22, 97–136
analysis methods, 106–114
building bioclimatic chart, 106–108
comfort triangle chart, 114
digital tools, 111–114

Givoni's building bioclimatic chart, 108–111
causes of climate differences, 97
classification systems
IECC, 103–104
Köppen, 100–101, 103
Köppen-Geiger, 101
climate zones, 100–103
climate zones and energy codes, 103–106
emissions modeling, 39, 40, 129–136
assumptions for simulations, 130–131
construction, 133
four climate zones, 134–136
operations, 130–132
transport, 134
waste, 133
water supply related, 133–134
passive cooling, 222–223
green cooling model, 241–242
nocturnal ventilation, 229
passive heating design strategies, effects on emissions, 284–288
radiation impact on surfaces, 42, 44, 45, 161
and radiative heat transfer, 259
solar radiation impact, 161
solar radiation modeling, 200
variables affecting architecture, 97–100, 101, 102
ventilation
building shape and openings, 219–220
types of, 215
vernacular architecture, 114–129
cold climates, 128–129
hot dry climates, 119–123
hot humid climates, 115–118, 119
temperate climates, 124–127, 128
weather, defined, 97
Climate analysis
emissions reduction, 23
methods, 106–114
studio projects
advanced, 38
beginning year, 27, 28, 29
Climate analysis software, 100, 101, 106, 111–112
Climate and House Design (Koenigsberger, Mahoney, and Evans), 106
Climate change, 3–6, 22
Climate Consultant, 100
Olgyay's timetable of bioclimatic needs, 112, 113
Pomona climate plot, 238
studio projects, beginning year, 27, 28
temperature information for sundials, 157, 160
weather data, 157

Index

windows and shading design, 44
    south facing window, 150, 153
    sundials, 157, 160
  wind wheel, 112, 113, 114
Climate zones, 100–103
  and emissions, 28
    geographic distribution of, 39
    modeling, 134–136
  and energy codes, 103–106
  studio projects, advanced, 34–35
Clothing, 85–86, 88, 93, 95, 210, 224, 225
Cloud cover
  climate analysis, 97, 98
  green cooling model, 241
  longwave radiation losses, 197
  and passive heating, 281
  radiative heat exchange, 258, 259
  roof pond radiant loss, 247–248
  and solar radiation, 271
Cluster analysis, hierarchical, 103
Coatings, glass, 186, 205
Code compliance
  building classification, 20
  design process, 19
Cold climates
  classification systems, 102
  conductive heat exchange, form factor and, 189–190
  design strategy effects on emissions, 286, 287
  earth tube cooling systems, 265–266
  emissions simulations, 131
  heat balance of buildings, 171
  operations emissions, 134–136
  passive heating systems; *See* Heating, passive
  solar gains, 195
  solar radiation effects, 194
  vernacular architecture, 128–129
  window types for, 208
Cold energy transfer, 259, 261
Colonial architecture
  corporate housing in hot humid climates, 117–118
  temperate climates, 124–125, 126, 127
Colors, surface, 193
  cold climates, 194
  and conductive heat transfer, 183
  green cooling model, 238, 241
  and radiative heat transfer, 192, 202, 203
  solar daylighting, 273
  water walls, 279
Combustion, 9, 57
Comfort, thermal; *See* Thermal comfort
Comfort chart, FSEC, 224
Comfort low/comfort high values, smart ventilation systems, 214

Comfort models
  adaptive, 90, 92–95, 96
  climate analysis tools using, 112
  physiological, 89–90, 91
Comfort scale, 88
Comfort ventilation, 221, 222, 223, 224–226, 227
Comfort zone
  comfort ventilation and, 223–224
  Givoni's, 110
  passive cooling, 222
  smart ventilation model system, 232–233, 235
  thermal, 94
Community well-being, neighborhood design, 40
Compatibility, design tools, 18
Complex envelopes, solar radiation modeling tool, 200, 201
Composting, 24–25, 57
Computational fluid dynamics (CFD), 53, 54
Computer tools; *See* Digital tools/software
Concrete
  absorptivity and emissivity, 193, 203
  floor materials, 284, 285
  heat capacity, 180
  thermal storage, 276
Concrete pavers, 44, 238
Concrete slabs, 284
Condensation
  adiabatic, 243
  latent heat, phase changes, 173–174
Conductance, 175
  heat balance equations, 170
  and resistance, 176
  transmittance (U-value), 178–179
Conduction, thermal, 165
Conductive heat exchange, 165, 174–190
  decrement factor, 179–180, 183
  design strategies, 180, 183–190
    air spaces, 185
    energy storage capacity of materials, 187
    high-density and high thermal capacity materials, 187–188
    insulating materials, 184–185
    rate of heat flow, Fourier's equation, 183–184
    surface area reduction, 188–189
    surface thermal resistance, walls and roofs, 185
    thickness of architectural elements, 186
    windows, thermal resistance of, 185–186
  earth coupling, ground/soil as heat sink, 263–267
  glazing to surface ratios and, 209
  heat balance equations, 166, 167
  heat loss in cold climates, 195

material characteristics, 174–180
  conductance, 175
  conductivity, 175, 176
  heat capacity and specific heat capacity, 178–179, 181
  resistance, 176–177, 180
  transmittance (U-value), 177–178, 179
passive cooling
  earth coupling, ground/soil as heat sink, 263–267
  heat loss prevention, 271
  night ventilation considerations, 230
passive heating
  heat loss, 269
  thermal storage systems, 276–277
radiation impact on surfaces, 42, 161
shading, 144
solar position and
  orientation of building and, 163
  solar radiation impact, 161
solar radiation, 194–195
  opaque components, 197
  solar protection, types of, 198
  thermal effect of, 196
thermal comfort, 82, 83, 102
time lag, 179, 180, 182
Conductivity, 175, 176
  and rate of heat flow, 183
  and resistance, 176
Conservation Fund, 12
Construction
  HMC Architects flex school building, 70
  life-cycle of buildings, 19
  studio projects, advanced, 56
  zero-waste, 56
Construction emissions, 9–10, 22
  carbon counting tools, recommended, 15
  climate effects simulations, 133, 134
  design process, 23–24
  design process roadmap, 20, 21
  HMC Architects flex school building, 72
  studio projects, beginning year, 27
Construction materials
  emissions from, 270
  low-energy envelope design exercise, 42
  operations emissions, 9
Consumption, reduction of, 19
Control systems
  night ventilation, 230
  smart ventilation model, 231, 232, 233
  thermal comfort, 94, 95, 96
Convective heat exchange, 165, 210–220
  air movement and infiltration, 211–213, 214
  air spaces, 177, 178, 185
  building heat balance, 168

controlling air exchange by opening building, 214–220
  building shape and, 219–220
  effects of skin and internal elements on airflow, 215–219
  surroundings and, 220
  windows and, 219
controlling air exchange by sealing building, 214
definition of, 210–211
passive cooling; *See* Ventilation
passive heating
  heat loss prevention, 269, 270, 271
  isolated gain systems, 282–283
  roof ponds, 281
  Trombe walls, 279
  water walls, 279
prevention of overheating/overcooling, 168, 169
radiant cooling system, 259
shading, 144
sol-air temperature, 196
solar energy dissipation, 195, 204
solar protection, types of, 198
surface resistance, 185
surface roughness and, 201
thermal comfort, 82, 83, 101–102
Cook, J., 219, 222, 224, 228, 258
Cook, Ryan, 46
Cooling
  adiabatic, 83
  climate variables, 98
    building bioclimatic charts, 109
    vernacular architecture; *See* Vernacular architecture
  low-energy envelope design exercise, 41, 42, 43, 44
  operations emissions, 9
  percentage of emissions in geographic subareas, 39
  solar radiation, 161
    shading, 144, 150, 154
    window selection, 202–206, 207
Cooling, passive, 22
  applicability, 267, 268
  bioclimatic design, 221
  classification of, 221–223
  convective heat exchange
    closing building, 214
    controlling air exchange, 213–214
    opening building, 214–220
  definition of, 221
  design strategy effects on emissions, 284–288
  earth coupling, ground/soil as heat sink, 263–267

Index

evaporative cooling, 242–257, 258
form factor, 189
HMC Architects designs
    Cal Poly Pomona building, 65, 74
    CSUMB classroom building, 65
    flex school building, 65, 66–72
radiant cooling, 258–263
shading, 144
studio projects, advanced
    direct gain, 54
    low-energy envelope design exercise, 41, 43
    natural ventilation, 54
    solar and site analysis, 39
studio projects, beginning year, 28, 31
surface roughness and, 201
temperature reduction using high-density and high-thermal capacity materials, 187–188
ventilation, 223–242
    applicability, 267–268
    comfort, 224–226, 227
    green cooling, 236–242
    nocturnal, 226, 227, 228–237
    smart ventilation, 231–235
    smart ventilation, shading effects on, 235–236, 237
vernacular architecture; *See* Vernacular architecture
Cooling degree day, defined, 99
Cooling towers, 42, 244, 245
CORAZON project, 34, 35, 36
Corporate housing, hot humid climates, 117–118
Cosine effect, 137, 138, 194
Courtyards, 65, 121–123, 124
Cross-ventilation
    convective heat transfer, 214
    effects of skin and internal elements on airflow, 215–219
    neighborhood design, 41
Curtains, 146, 186, 210, 235
Cyclic process, design as, 17, 18

D

Dampers
    isolated gain systems, 282–283
    smart ventilation model system, 232, 234
Danish building codes, 106
Dates and times, Climate Consultant windows design, 44
Dayag, Ryan, 31
Daylength, seasonal changes, 137
Daylighting/daylight analysis, 22
    climate analysis, meteorological data, 98
    design process roadmap, 21

HMC Architects CSUMB classroom building, 65, 73
HMC Architects flex school building, 65, 68
operations and energy emissions reduction, 23
solar buildings, 272, 273
studio projects, advanced, 34, 49–50, 51, 52, 53
studio projects, beginning year, 25–26, 27, 28, 30, 31, 32, 33
windows and shading, 44
windows and shading design, 47
Day-night cycles; *See also* Temperature swings
    climate analysis, 98
        building bioclimatic charts, 107–108
        hot dry climates, 102–103, 119, 120, 121
    passive cooling
        radiative heat exchange, 260–261
        roof pond performance, 247–249
        ventilation control, 219–220
        ventilative, 226, 227, 228–231
    passive heating
        direct gain, 273–274
        isolated gain systems, 283–284
        roof ponds, 281, 282
        Trombe and mass walls, 276, 277
        water walls, 280
    shading systems, movable, 198, 199
    sun position, 139, 140
Declination, sun, 137, 139, 140, 201
Decrement factor, conductive heat exchange, 179–180, 183
DEFRA, 14
Degree day, defined, 99
Dehumidification, 78, 83, 214
Demolition, 19
Density, air, 212
Density, material
    and conductive heat transfer, 174, 175, 185
    energy storage capacity of materials, 187
    heat capacity and specific heat capacity, 179
    heat exchange
        decrement factor, 183
        time lag, 182
    insulating materials, 185
    temperature reduction using high-density and high-thermal capacity materials, 187–188
Design Builder, 132
Developed versus undeveloped countries, thermal comfort in, 109–111
Diaz, Roderigo, 47
Diffuse light/radiation, 30, 184, 194, 195, 202
    analysis of, 162, 163
    climate and, 100
    energy flow design, 166, 167

glazing to surface ratios and, 209
shading, 121
solar daylighting, 51, 63, 68, 272, 273
Digital tools/software; *See also specific programs (Ecotect, etc.)*
climate analysis, 100, 101, 106, 111–112
climate effects simulations, 132
solar radiation on building envelopes, 200, 201
studio projects, advanced, 29
studio projects, beginning year, 26, 27–28
Tijuana low-cost sustainable housing prototype project, 34
Dimensions; *See* Geometry
Direct emissions, 9, 10
Direct evaporative cooling systems, 243–246
Direct gain systems
passive heating, 273–275
passive solar heating studio design, 54
Trombe wall heating, 279
Direct radiation, solar, 194, 195, 196, 202
energy flow design, 166, 167
glazing to surface ratios and, 209
Dirty buildings, 19–20
Diurnal cycles; *See* Day-night cycles
Domestic water heating; *See* Water heating
Double glazing; *See* Multilayer glazing
Double-play system, 263
Drainage systems, 97
Drip irrigation, 56, 67, 73
Dry-bulb temperature
climate variables, 97, 98
building bioclimatic charts, 108
and convective heat transfer, 101
comfort chart, 225
operative temperature, 86, 88
passive cooling, 268
psychrometric charts, 75–78, 79–83; *See also* Psychrometric charts
and sensible heat, 77
and thermal neutrality temperature, 90, 92
Dry climates, 103; *See also* Hot dry climates
Dry fixtures, 57, 66
Duong, John, 41

E

Earth, thermal equilibrium, 258
Earth coupling, ground/soil as heat sink, 221, 222, 263–267, 268
Earth tubes, 265–267
East facing elevation, solar radiation, 162–163, 208
Eaves, 220
Eco-Calculator, 55
Ecological footprint, 10–11
Ecotect

air flow/air movement analysis, 53
illuminance and luminance levels, 49–50
incident radiation calculation, 162
radiation impact on surfaces, 42
solar and site analysis, 39
solar geometry
site analysis, 159
sun path diagrams, 142, 144
solar radiation on building envelopes, 200
studio projects, advanced, 34
studio projects, beginning year, 27, 28
Weather Tool, 100, 101
windows and shading design, 44
Efficiency
California standards, 104, 105
design strategy effects on emissions, 284, 285, 287, 288
emissions reduction, 8
solar heating collector, 54
eGRID database, 132
Electric Power Pollution Calculator, 12, 13
Electromagnetic radiation, 165, 174, 190
Emissions, 1–15
buildings and, 7–10
construction emissions, 9–10
operations emissions, 9
waste, emissions from, 10
water, emissions from, 10
carbon counting tools, 10–14
comparisons of emissions from natural gas and electricity, 12–14
recommended, 14, 15
selection of, 11, 12
climate related effects modeling, 129–136
assumptions for simulations, 130–131
construction, 133
four climate zones, 134–136
operations, 130–132
transport, 134
waste, 133
water supply related, 133–134
design goals, 18–20
effects of design strategies, 284–288
effects on climate change, 3–6
greenhouse effect, 1, 2
types, sources, and measurement of, 1–2
Emissivity, 202
factors affecting solar radiation, 191
insulating materials, 185
radiative heat exchange, 201–202, 203
and resistance, 177
solar radiation, 192
surface roughness and, 201
Emittance
low-E coated glass, 205, 207
solar radiation, 193

Index

Emitter-sequester classification of buildings, 19–20
EN15251, 94
Enea, 116
Energy
  defined, 165
  operations emissions, 9
  storage capacity of materials, 187; *See also* Mass, structural/building; Storage, energy; Thermal mass
Energy Conservation through Building Design (Milne and Givoni), 109
Energy Design Resources, 36
Energy efficiency; *See* Efficiency
Energy flows, 165–171; *See also* Heat exchange principles
  design for control of, 166–168, 169
  heat balance, building, 170–171
  operations emissions simulations, 130
Energy generation
  climate variables, 98
  design process roadmap, 21
  emissions from
    baseline emissions determination, 19
    design process, 23
    recycling and, 60
    studio projects, advanced, 37
  photovoltaic systems; *See* Solar energy/photovoltaic systems
  from waste, 22, 24, 57
Energy modeling tools, 208; *See also* Digital tools/software; *specific programs* (Ecotect, etc.)
  passive heating, direct gain, 275
  shading system inclusion in, 157
  studio projects, advanced, 36–38
  studio projects, beginning year, 26–27
EnergyPlus, Design Builder, 132
Energy Star, 103
Energy Star appliances, 44
Energy Star Home Energy Yardstick, 12
Energy Star Target Finder, 12
Entenza, John, 125
Enthalpy, 75, 77, 78; *See also* Psychrometric charts
Envelope, building
  and air movement, 215
  design process roadmap, 21
  design strategy effects on emissions, 284–285
  energy flow design, 166, 167
  heat balance equations, 170
  heat exchanges through, 165–171
    concepts, 165
    conductive, 167, 174–190; *See also* Conductive heat exchange
    convective, 210–220; *See also* Convective heat exchange

    design for control of energy flows, 166–168
    heat balance, building, 170–171
    mechanisms of heat flow, 165, 166, 167
    overheating and overcooling, strategies to prevent, 168, 169
  operations emissions simulations, 130, 131
  solar heating, direct gain, 274–275
  solar radiation modeling, 200, 201
  studio projects, advanced, 41, 42, 43, 44
  studio projects, beginning year, 25, 26, 27, 28
  surface area reduction, 188–190
Environmental variables, and thermal comfort, 82, 84–85
Environment-building interactions, greenhouse gas emissions, 8
Eolic orientation of building, 209
EPA Personal Emissions Calculator, 12, 13, 15
EPA Waste Reduction Model WARM, 12, 15
EPW file format, 99, 111
Equator, 138
Equatorial orientation; *See* Orientation
Equatorial plane, solar declination, 139
Equilibrium, thermal, 165, 166–167, 170–171
Equinoxes, 140
Equipment, heat generation from; *See* Internal gain
Europe, natural ventilation strategies in, 230
European standards
  energy efficiency, 104, 106
  thermal comfort, 94
Evaluation, post-occupancy (POE), 19
Evaluation phase, design process, 18
Evans, M., 106, 114
Evaporation, latent heat of, 172–173
Evaporative cooling/heat loss
  air movement and, 84–85
  air speed and, 215
  building bioclimatic charts, 109
  building heat balance, 168
  energy flow design, 166, 167
  green roofs, 237
  heat balance equations, 167
  human body, 80, 81, 82, 83
  latent heat, 172–173
  low-energy envelope design exercise, 43
  outdoor space cooling, 222
  prevention of overheating/overcooling, 168, 169
  and psychometric chart, 78
  soil cooling, 267
  thermal comfort, 101, 102
Evaporative cooling systems, 242–257, 258, 268
  direct, 243–246
  indirect, 244, 246–257, 258
    Cal Poly Pomona modular roof pond, 255–256, 257, 258

Cal Poly Pomona smart roof pond with floating insulation, 254–255
roof ponds in hot humid climates, 253–254
UCLA roof pond, Givoni-La Roche, 250–253
University of Nevada roof pond, 256–257
physical principles, 242
Evapotranspiration, 66, 67, 237; *See also* Plants, use of
EvSurf, 200, 201
Exterior elements; *See also* Envelope, building
shading on windows, 44, 51
shadow angle protractor, 149
solar protection for, 197, 198–200
External shading, 209, 210

F

Fans, 42, 214, 226, 228
Fahrenheit scale, 165
Fenestration; *See* Window systems
Fertilizers, sustainable, 67
File compatibility, design tools, 18
File formats, weather data, 99, 111
Fins, 199–200
and air movement, 219
roof pond, 257
shadow angle protractor, 149, 159
Flex school building, 63, 65–72
Floorplan, open, 272
Floors
emissions effects of design strategies, 284, 285
glazing to surface ratios, 209, 217, 232, 241, 254, 275, 283
passive heating
direct gain, 53, 273–274
heat loss, routes of, 269, 271
thermal mass, 263, 273
temperature, 81
vernacular architecture
courtyard, 121, 122–123
traditional buildings, 116
Fluid dynamics
computational (CFD), 53, 54
convection, 210
Flux rate, conductive heat flow, 183
Folding panels, 210
Forced airflow, energy transfer, 259
Forced convection, 211
Forced ventilation, 211, 214, 226, 228, 231–236
Form factor; *See* Geometry
Fountains, 243
Fourier's equation, 183–184
Fox, Michael, 28
Freeze protection, solar hot water systems, 54

Freezing (latent heat of fusion), 171–172
Fuel cells, 57
Fuels
emission reduction, 8
landfill gas, 57
operations emissions, 9
Furnaces, 9
Fusion, latent heat of, 171–172

G

Garcia, Marcos, 41
Gas-liquid phase change, 172–173
Gayomali, John, 40
GEIC Calculator, 12
Generation/reduction phases, design process, 17, 18
Geographic distribution of emissions, 38–39, 40
Geographic location
and solar radiation; *See* Latitude
temperate climates, 123
Geometry; *See also* Thickness, material
and air movement, 215, 218–220
building surface area reduction, 188–190
cold climate vernacular buildings, 129
and conductive heat exchange, 167, 184, 186
conductive heat transfer, 167, 184
control of heat loss, 269
CSUMB classroom building, 63, 65, 73
glazing to surface ratios; *See* Glazing systems, glazing to surface ratios
solar radiation modeling tool, 200, 201
Trombe and mass walls, 277–278
wind, designing for, 220
Givoni, B, 107, 108–111, 217, 219, 221, 228, 230, 241, 243, 245, 249, 250, 253, 254, 259, 261
Givoni-La Roche roof pond, 249, 250–253
Glare, 28, 49
Glass
absorptivity and emissivity, 203
radiation wavelength and transparency, 192, 194
solar radiation transmission, 203
Glazing systems, 186; *See also* Window systems
glazing to surface ratios, 208, 217, 232, 254, 275, 283
green cooling model, 241
radiative heat exchange, 209
solar collecting
direct gain systems, 273–275
indirect gain systems, 275–284
Global solar energy, 209
Global warming, 1
Global warming potential (GWP), 2
Globe thermometer, 191

Index

Goals
  design process roadmap, 21, 22
  studio projects, beginning year, 27
Gonzalez, Emmanuele, 43, 47
Gonzalez, G., 165, 166, 198
Gray water reuse, 57, 66
Green cooling, 236–242
Green footstep, 14
Greenhouse effect, glass, 194, 202
Green Kit, 28–29, 33, 53
Green roof
  green cooling, 236–242
  HMC Architects flex school building, 67, 71
  low-energy envelope design exercise, 43
  solar protection, types of, 198
Green screen, 66, 67, 71
Green sustainable building category, 20
Ground heat conduction, 271
  energy flow design, 166, 167
  passive cooling, 221, 263–267, 268
    direct contact, 264–265
    heat exchangers, 265, 267
    surface and subsurface cooling, 267
Ground reflection, 166, 167, 207
Grouping of buildings; *See* Massing of buildings
Gulloti, Brandon, 40

H

Haciendas, 124
Hansanuwat, Ryan, 34, 41, 239
Hanson, C.E., 3, 7
Harold Hays Skytherm System, 249, 256–257, 259–260
Hatos, 120–121
Hawaii, 118, 119
Hay, Harold, Skytherm System, 249, 256–257, 259–260
Heat balance, physiological; *See* Physiology/heat balance
Heat (thermal) balance equations, 166–167, 170–171
Heat (thermal) capacity and specific heat capacity, 178–179, 181
  common materials per unit mass and per unit volume, 180
  high-density and high thermal capacity materials, 187–188
  smart ventilation model system, 233
Heat content, enthalpy, 77, 81
Heat exchange; *See also specific modes of heat exchange* (conductive, convective, radiative)
  physiological comfort model, 89–90
  radiation impact on surfaces, 42
  shading, 144
  vernacular architecture; *See* Vernacular architecture

Heat exchange principles, 165–220
  envelope, energy exchanges through, 165–171
    concepts, 165
    design for control of energy flows, 166–168
    heat balance, building, 170–171
    mechanisms of heat flow, 165, 166, 167
    overheating and overcooling, strategies to prevent, 168, 169
  forms of heat, 171–174
    latent heat, 171–174
    radiant heat, 174
    sensible heat, 171
  mechanisms of heat exchange, 165–171, 174–220
    conduction, 174–190; *See also* Conductive heat exchange
    convection, 210–220; *See also* Convective heat exchange
    radiation, 190–209, 210; *See also* Radiative heat exchange
  passive cooling; *See* Cooling, passive
Heat exchangers
  earth-to-air, 265–267
  evaporative cooling, 242, 245, 246
Heat flow, body, 79
Heat flow rate, convective heat exchange, 213
Heat gain
  heat balance equations, 166–167
  heat balance of buildings, 171
  radiative, 269; *See also* Solar gain; Solar heating
Heating
  adiabatic, 83
  building bioclimatic charts, 109
  climate variables, 98
  cold climate vernacular buildings, 128, 129
  degree days, 99
  low-energy envelope design exercise, 41, 42, 43, 44
  operations emissions, 9
  percentage of emissions in geographic subareas, 39
  solar radiation
    shading, 144, 150, 154
    window selection, 202–206, 207
  solar radiation impact, 161
  vernacular architecture; *See* Vernacular architecture
  window selection, 203–206, 207
Heating, passive, 22, 222, 258, 269–288
  applicability, 269, 270
  building bioclimatic charts, 109
  California State University classroom building design, 65
  control of heat loss, 269–270, 271
  direct gain systems, 273–275

emissions effects of design strategies, 284–288
form factor, 189–190
indirect gain systems, 275–284
    roof ponds, 281–282
    sun rooms/sunspaces, 282–284
    Trombe and mass walls, 276–279
    water walls, 279–281
shading, 144
solar, 270–273
studio projects, advanced
    direct solar gain, 54
    low-energy envelope design exercise, 41, 43
    solar and site analysis, 39
studio projects, beginning year, 28, 31
temperature reduction using high-density and high-thermal capacity materials, 187–188
Zomeworks Double-play system, 263
Heating degree day, 99
Heat island effect, 67, 71
Heat loss
    building orientation and, 272
    control of, 269–270, 271
    heat balance equations, 166–167
Heat radiant floor, 43
Heat sinks; *See also* Mass, structural/building; Storage, energy; Thermal mass
    energy storage capacity of materials, 187
    latent heat, 171–174
    natural ventilation, 54
    passive cooling, 221, 222, 268
        green cooling model, 237, 239
        nocturnal ventilation, 230
        roof ponds; *See* Roof ponds
    structural mass as, 226, 227, 228
Heat transfer; *See* Heat exchange; *specific modes of heat exchange* (conductive, convective, radiative)
HEED (Home Energy Efficient Design)
    climate effects simulations, 132
    emission area applications, 12
    night ventilation, 230
    recommended carbon counting tools, 15
    studio projects, advanced, 41, 55
    studio projects, beginning year, 27, 28
Height; *See* Geometry
Hernandez, Alexandra, 52, 53
Hierarchical cluster analysis, 103
High-density and high thermal capacity materials, 187–188
High-performance building, 20
HMC Architects, 14, 60, 61, 63, 65–73
    California State University classroom building design, 63, 65, 73
    flex school building, 63, 65–72
    surface analysis of direct radiation, 162, 163

Home Energy Efficient Design; *See* HEED
Home Energy Saver Calculator, 12, 39
Horizontal building elements, shading design, 154
Horizontal shadow angle (HSA)
    shading design, 146, 147
        southeast facing window, 155–156
        south facing window, 151
        shadow angle protractor, 149, 150
Horizontal sun path diagram, 140
    shading design, 151, 153
    site analysis, 159
    solar charts, 142, 143, 144
Horizontal toplights, 209
Hot climates
    convective heat transfer, 214
    passive cooling; *See also* Cooling, passive
        nocturnal ventilation, 231
        radiative heat exchange, 260–261
    window types for, 208
Hot dry climates
    classification systems, 102
    design strategy effects on emissions, 284–285, 286, 288
    emissions simulations, 131
    evaporative cooling, 242, 244
    form factor, 189
    longwave radiation losses, 197
    operations emissions, 134–136
    passive cooling
        nocturnal ventilation, 228
        roof ponds; *See* Roof ponds
    radiation impact on surfaces, 44, 45
    studio projects, advanced, 34–35
    ventilation
        building shape and openings, 219–220
        types of, 215
        wind patterns, 54
    vernacular architecture, 119–123
Hot humid climates
    classification systems, 102
    design strategy effects on emissions, 286
    diffuse light in, 162
    emissions simulations, 131
    evaporative cooling, 242
    form factor, 189
    longwave radiation losses, 197
    operations emissions, 134–136
    passive cooling
        comfort ventilation, 225, 226
        roof ponds, 253–254
    ventilation
        building shape and openings, 219–220
        types of, 215
    vernacular architecture, 115–118, 119
Hot water; *See* Water heating
Hour angle, 137, 139, 140, 201

Index

Hourly data
    climate data sets, 99
    timetable of bioclimatic needs, 112, 113
Housing prototypes, studio projects, 34–35, 36
HSA; *See* Horizontal shadow angle
Human activities, greenhouse gas emissions; *See* Emissions
Humid climates; *See also* Hot humid climates
    building bioclimatic charts, 108
    IECC climate zones, 103
Humidity; *See also* Absolute humidity; Relative humidity
    climate analysis
        climate types, 103
        meteorological data, 97, 98
    comfort chart, 225
    heat balance, human body, 85
    indoor sources of, 214
    operative, 91
    and passive cooling, 268
        evaporative, 242
        nocturnal ventilation, 228
    psychrometric charts, 75–78, 79–83; *See also* Psychrometric charts
    and thermal comfort, 75
        developed versus undeveloped countries, 109–111
    water walls and, 281
Humidity ratio, 91
Hybrid ventilation strategies, 230
Hydrofluorocarbons, 1, 2

I

IAQ; *See* Indoor air quality
IECC climate zones, 103–104
Illuminance, 26, 28, 49
Illuminance meters, 26
Incidence, solar radiation, 138
Incident radiation, 44, 194
Incinerator emissions, 60
Inclination, surface resistance, 185
Inconvenient Truth, 12
Indirect evaporative cooling; *See* Evaporative cooling systems, indirect
Indirect gain systems, passive heating, 275–284
    roof ponds, 281–282
    sun rooms/sunspaces, 282–284
    Trombe and mass walls, 276–279
    water walls, 279–281
Indirect GHG emissions, 9, 10
Indoor air quality (IAQ), 22
    construction emissions reduction, 24
    design process roadmap, 21
    functions of ventilation, 223
    heat loss prevention and, 270
    sealed building, 214

Indoor-outdoor temperature differences; *See* Temperature difference, indoor-outdoor
Industrial process emissions, 1
Inertia, thermal; *See* Thermal mass
Infiltration, 210
    control of heat loss, 269–270
    convective heat exchange, 211–213, 214
    heat gain, 230, 231
    heat loss via, 269
    heat transfer, 167
    prevention of overheating/overcooling, 168, 169
    shading, 144
    smart ventilation model system, 234
Information exchange capability, design tools, 18
Infrared, radiant heat, 174, 190
Insolation, 44, 47, 48; *See also* Solar radiation
Insulation, 156
    climate variables, 99
    conductive heat exchange
        design approaches, 184–185
        form factor and, 190
        thermal mass as insulator, 181
    and convective heat transfer, 210
    design strategy effects on emissions, 286, 287, 288
    passive cooling
        green cooling model, 236–237, 239, 240, 241
        nocturnal ventilation, 226, 227, 228, 229, 230
        radiative heat exchange, 261
        smart roof pond with insulation, 254–255, 256
        smart ventilation model system, 231–232
        UCLA system, 261–262
    passive heating
        control of heat loss, 269
        design strategy effects on emissions, 284–285
        direct gain, 274–275
        solar systems, 271
    structural insulated panels (SIP), 42, 56, 63, 284
    and window performance, 210
Integrated design process, 36–38, 60–61, 62, 63, 64
Intensity, solar radiation, 194
    seasonal changes, 137
    sol-air temperature, 196
Interdisciplinary team, 38
Intergovernmental Panel on Climate Change (IPCC), 3–4, 5, 6, 7, 8
Interiors
    air flow; *See* Air flow/movement
    energy simulation tool exercises, 36, 37

heat sources; *See* Internal gain
studio projects, beginning year, 25, 26
Interior shading devices, 210
Internal gain
  building heat balance, 167–168, 169
  heat balance equations, 166, 167
  night ventilation, 226, 230
  prevention of overheating/overcooling, 169
Internal heat sources, energy flow design, 166, 167
International Energy Conservation Code (IECC) climate zones, 103–104
International system, solar position angles, 139, 141
International System (SI) measurement units, 165
IPCC (Intergovernmental Panel on Climate Change), 3–4, 5, 6, 7, 8
Irradiance, 97
Irradiation, 44, 47–49, 50
Irrigation, 56, 67, 73
Isothermal heat storage, water wall, 280

J

Joule (J), 165, 171

K

Kaas, Kellene, 30
Kalwall, 206
Kelvin (K), 165
Knowledge areas
  design process, 20, 21, 22, 23
  design process roadmap, 20, 21
  studio projects, beginning year, 27
Köppen climate classification system, 100–101, 103
Köppen-Geiger classification system, 101

L

Labs, K., 168
Landfills, 57, 60
Landscaping, 21, 22, 24, 220
La Roche, P., 11, 18, 25, 28, 34, 116, 118, 126, 165, 166, 198, 200, 214, 229, 231, 233, 236, 237, 239, 249, 250, 254, 255, 256
Latent heat, 77, 171–174, 222, 242; *See also* Evaporative cooling systems
Latin American colonial architecture, 124–125, 126, 127
Latitude
  passive solar building orientation, 271
  radiation impact on surfaces, 42
  roof pond limitations, 281
  solar gain factor, 202
  solar geometry
    altitude of sun, 139
    orientation of building, 163
    position of sun, 140
    radiation impact on surfaces, 161
    shading design, south facing window, 151
    and solar irradiation, 137
    sundials, 157, 160
    sun path diagrams, 142, 143, 144
  solar radiation
    modeling, 200, 201
    orientation of building, 208–209
    solar protection systems, 198–199
    temperate climates, 123
Layered air spaces, insulation, 210
Layers, glass, 186; *See also* Multilayer glazing
Life-cycle analysis, carbon footprint, 10, 19, 55–56, 133
Liggett, Robin, 18, 100
Lighting; *See also* Daylighting/daylight analysis
  climate variables, 98
  energy flow design, 166, 167
  interior arrangements with solar heating, 272
  percentage of emissions in geographic subareas, 39
Light intensity, seasonal changes; *See* Seasons
Light shelves, 50–51, 68
Light-to-solar gain (LSG) ratio, 204, 208
Lime whitewash, 192
Liquids, phase change materials, 173–174
Liquid-solid-gas phase changes, 171–174
Live Neutral, 12, 13
Location
  and climate data sets, 99
  local conditions, and optimum orientation, 208
  site analysis, sun path diagrams, 158–159
  temperate climates, 123
Locke, Aaron, 61, 62
Longwave radiation
  building heat balance, 169
  energy flow design, 166, 167
  heat balance equations, 167
  prevention of overheating/overcooling, 168, 169
  radiative heat exchange, 259
  sol-air temperature correction factor, 197
  solar energy dissipation, 195, 198
  windows
    with reflective glass, 206
    single-pane, 204
Los Angeles Case Study Houses, 125–126, 128
Los Angeles Unified School District project, 63, 65–72
Louvers, 146, 210, 235
  Calearth Ecodome cooling system, 258
  exterior shading devices, 145
  low-energy envelope design exercise, 42, 44

Index

radiative heat exchange
   roof ponds, 249, 256, 257
Low-carbon materials, 24
Low-E coatings, 205, 207, 208
Low-flow fixtures, 56
LSG (light-to-solar gain) ratio, 204, 208
Luminance levels, daylighting, 49, 50, 52
Lyle Center for Regenerative Studies, 57
   green cooling model, 236–242
   roof pond, 254–255, 256

M

Machado, M., 116, 125, 165, 166, 198, 253, 229
Maracaibo, Venezuela, 118, 253–254, 260–261
Marine climates, 103
Martin, Ashi, 36
Masonry, 276
Mass
   air, convective heat exchange, 211
   energy storage capacity of materials, 187
   sensible heat, 171, 172
Mass, structural/building; *See also* Storage, energy; Thermal mass
   architectural elements, and conductive heat exchange, 186
   building bioclimatic charts, 110
   and daily temperature swing, 120, 121
   decrement factor, 180
   heat capacity, 179; *See also* Storage, passive heating systems; Thermal mass
   Olgyay building bioclimatic chart and, 107–108
   and passive cooling, 223
      computer programs for modeling, 229–230
      nocturnal ventilation, 226, 227, 228, 229, 230
      radiative heat exchange, 261
      smart ventilation model system, 234
      UCLA system, 262–263
      ventilation types, 215
Massing of buildings; *See* Surroundings, buildings and objects
Mass walls, passive solar systems, 276–279
Material characteristics
   conductive heat exchange
      conductance, 175
      conductivity, 175, 176
      energy storage capacity, 187
      heat capacity and specific heat capacity, 178–179, 181
      high-density and high thermal capacity materials, 187–188
      insulating materials, 184–185
      resistance, 176–177, 180
      surface thermal resistance, walls and roofs, 185
      transmittance (U-value), 177–178, 179
   radiative heat exchange
      absorptance and reflectance, 191, 192
      absorptivity and emissivity, 193, 203
      exterior surfaces, 201–202, 203
Material flows, construction emissions reduction, 24
Materials, 22
   absorptivity and emissivity, 203
   design process roadmap, 21
   emissions from, 8, 270
   HMC Architects CSUMB classroom building, 73
   HMC Architects flex school building, 70
   phase change (PCM), 173–174
   studio projects, advanced, 37, 55, 56
   Tijuana low-cost sustainable housing prototype project, 34
   vernacular architecture; *See* Vernacular architecture
Material selection
   construction emissions reduction, 24
   radiative heat flows, 201–202, 203
Maya (modeling software), 200
Mayer, Serge, 46
McCoy, Ester, 125
Mean radiant temperature (MRT)
   comfort chart, 225
   passive cooling, 225, 268
   radiant exchange, 84, 191, 209
   thermal comfort, 84, 88, 89
Measurement units
   emission baseline, 19
   energy, 165
   enthalpy, 77
   greenhouse gas emissions, 2
   heat, 165
   temperature, 75
Mechanical systems
   design strategy effects on emissions, 284–285, 286, 287
   low-energy envelope design exercise, 41
Mechanical ventilation, thermal comfort, 93
Mediterranean climate, 123, 124
Medrano, Jorge, 55, 60
Melting (latent heat of fusion), 171–172
Mensonides, Calvin, 33
Metabolic heat; *See* Physiology/heat balance
Metal roofs, 260–261
Metals, 180, 191, 203
Meteonorm, 111
Meteorological/weather data, 98–99
   climate analysis software and, 100, 101, 106
   Climate Consultant, 157
   digital climate analysis tools, 111–112

factors affecting design, 97, 98–99
file formats, 99, 111
shading system design, 156
Methane emissions
carbon counting tools, 12–14
global warming potential, 2
landfill gas collection and use, 57
recycling and, 60
sources of, 1
Mexico
GreenKit project, 29
vernacular architecture, 126
Milne, Murray, 28, 41, 100, 109, 110, 112, 114, 132, 214, 230, 231, 233, 236, 241
Modeling tools; *See also* Digital tools/software
studio projects, advanced, 34
studio projects, beginning year, 27, 28
Models of thermal comfort, 89–95
Modular design, 55, 70
Moisture content; *See also* Absolute humidity; Psychrometric charts
absolute humidity, 75, 79
building bioclimatic charts, 108
climate analysis, 101–102
climate variables, 98
and heat content, 78
and thermal comfort, 102
Mold and mildew, 225, 265
Monasteries, 124
Mountains, 129
Movable panels
insulating materials, 261
shading systems, movable, 199
UCLA radiant cooling system, 261–262
MRT; *See* Mean radiant temperature
Mulching, soil cooling, 267
Multilayer glazing (double and triple), 186
design strategy effects on emissions, 286, 287, 288
passive heating
direct gain systems, 274
sunspace, 283
Trobe wall performance, 279
window systems, 205, 206, 207, 208

N

Nanogel, 206
National Aeronautics and Space Administration (NASA), 3–4, 5
National Oceanic and Atmospheric Administration (NOAA), 103
National Renewable Energy Laboratory (NREL) model, 132
Native plant use; *See* Plants, use of
Natural convection, 210
Natural gas emissions; *See* Methane emissions

Natural ventilation, 51, 53, 54
passive cooling, 211
thermal comfort, 94
Nature Conservancy's Carbon Footprint Calculator, 12
NCARB Grand Prize, 34
Neighborhood design, 40–41
Neighborhood scale analysis
energy simulation tool exercises, 36, 37
studio projects, beginning year, 25, 26
Nelson, Phyllis, 28
Neutrality temperature
adaptive comfort model, 93–94
European standards, 95
New Orleans, Louisiana rainwater harvesting proposals, 57, 58, 59, 60
New Zealand climate data sets, 99
Ngo, Kenny, 30
Nightwall, 210
Nitrous oxide, 1, 2
Nocturnal cooling, 268
Cal Poly Pomona building, 74
design strategy effects on emissions, 288
green cooling model, 238–239, 241
ventilative, 221, 222, 226, 227, 228–237
Northern hemisphere, solar geometry, 137, 140
North facing slopes, 137
North pole, 137, 138
Nubosity, 98, 101–102

O

Obstructions; *See also* Surroundings, buildings and objects
Occupancy, interior arrangements with solar heating, 272
Occupants
carbon dioxide emissions, 270
heat generation from; *See* Internal gain
Occupation density, ventilation, 218
Oil company housing, hot humid climates, 117–118
Oil platform city, 36
Olgyay, V., 106–108
Olgyay's timetable of bioclimatic needs, 112, 113
Opacity/opaque elements, radiative heat exchange, 190, 191, 192, 194, 195–200
shading devices, 197
types of solar protection, 198–200
Open floor plans, 272
Open Graphics Library (OpenGL) modeler, 132
Openings, placement and size; *See also* Window systems
air movement, 215–220, 268
design with wind, 51, 53, 54
in hot dry climates, 121
passive cooling, 231, 268

Index

and base temperature, 99
passive heating
    heat loss prevention, 271
    solar, 272
    Trombe wall, 276
solar access analysis, 44
Operations
    cost reduction, flex school building, 71
    life-cycle of buildings, 19
    night ventilation efficiency, 230
Operations emissions, 9, 10, 22
    carbon counting tools, recommended, 15
    climate effects simulations, 130–132, 134–136
    design process, 20, 21, 23
    HMC Architects flex school building, 72
    studio projects, beginning year, 27
Operative temperature, 86, 88, 91
Orbit, earth revolution, 138
Orientation
    air inlets/outlets, 218–219
    glazing to surface ratios, 209
    radiation impact on surfaces, 42
    shading design; *See* Shading design
    solar gain factor, 202
    solar geometry, 162–163
    solar heating, 271, 272, 273, 274–275
    solar radiation modeling, 200
    solar radiation summers versus winters, 208
    vernacular architecture in temperate climates, 124
    wind, designing for, 220
Oscillation, temperature; *See* Temperature swings
Outcomes, design process roadmap, 20, 21, 22
Outdoor conditions
    building bioclimatic charts, 107–108
    passive heating, psychrometric charts, 269, 270
Outdoor space cooling, 222
Outdoor temperature
    Climate Consultant, 157
    climate variables, 99
    heat exchange; *See also* Temperature difference, indoor-outdoor
        convective, 210–220
        heat balance equations, 170–171
    and night ventilation, 229
    sol-air temperature, 196
    thermal comfort
        adaptive model, 94
        standards, 92–93, 95
        and thermal neutrality, 92
Overhang
    emissions reduction, 284
    exterior shading devices, 145
    low-energy envelope design exercise, 44
    shadow angle protractor, 149, 150
    shadow angles, 146–148
    window design
        southeast facing window, 155
        south facing window, 153, 154
Overheating/overcooling, 22
    design process roadmap, 21
    HMC Architects Cal Poly Pomona building, 74
    HMC Architects CSUMB classroom building, 65, 73
    operations and energy emissions reduction, 23
    strategies to prevent, 168, 169

P

Pacific Gas and Electric, 12, 13
Palm Springs, California project, 39, 40, 45
Palutikoff, J.P., 3, 7
Pamo Valley project, San Diego County, 35, 60–61, 62, 63, 64
Panama, 118, 119
Panels
    insulating materials; *See* Insulation
    photovoltaic systems; *See* Solar energy/photovoltaic systems
    shading devices, 145, 210
Parallel shading devices, 198, 199, 200
Paraujano house, 116, 117
Park, Nancy, 48, 49, 50, 51
Parry, M.L., 3, 7
Passive heating; *See* Heating, passive
Pavers, 44, 238
Perception of comfort, 80, 95–96
Perfluorocarbons, 1
Perpendicular solar protection, 198–200
Pervious surfaces, 74
Phase change
    evaporative cooling, 242
    latent heat, 171–174
Phases, design project, 19
Photovoltaic design; *See also* Solar energy/photovoltaic systems
    solar and site analysis, 39
    studio projects, advanced, 43, 44, 55
    studio projects, beginning year, 28
Phung, Alex, 36
Physical models, 25–26
Physiology/heat balance, 79–82, 83, 84–86, 87, 88, 93, 101, 268
    convective heat transfer, 214
    perception of comfort, 95–96
    thermal comfort models, 89–90, 91
Plants
    carbon storage and, 60
    climate analysis, site-specific conditions, 99

Plants, use of
  green cooling, 236–242
  HMC Architects CSUMB classroom building, 73
  HMC Architects flex school building, 66, 67, 71
  low-energy envelope design exercise, 43, 44
  solar protection, 198, 200
  studio projects, advanced, 43, 44, 56
Plaster, 193
PMV/PPD model, thermal comfort, 88, 89–90, 92
Polar facing elevations, 162, 199, 208, 272
Pool heating, solar, 54
Population density, 39, 40
Porous asphalt, 44
Post-occupancy evaluation (POE), 19
Power Profiler, EPA, 132
Precipitation, 97, 98, 99, 101
Predicted mean vote/PPD model, thermal comfort, 88, 89–90, 92, 112
Predictive value, computer models, 34
Pressure differences
  natural ventilation, 226
  wind-generated, 214, 215, 216, 217–218
Problem solving, 17
Professional practice, CNAD, 61, 63, 65–74
Profile angle (vertical shadow angle), 146
Protractor, solar, 149, 150, 151, 155–156, 158
Psychological adaptations, and thermal comfort, 80, 93
Psychrometric charts, 75–78, 79–83
  building bioclimatic charts, 108–109, 110, 111, 112
  evaporative cooling, 243
  passive cooling
    comfort ventilation, 224
    nocturnal ventilation, 229
  passive heating, 269, 270
  thermal comfort modeling, 90, 91
  windows and shading design, 44
Puretrust, 12, 13
PVWatts, 27, 28, 55
Pyranometer, 97

Q

Quirós, C., 109, 166, 198

R

Radiance (program), 27, 49–50
Radiant barrier, 185
Radiant cooling, 65, 258–263
Radiant energy flow from sun, 166, 167
Radiant floors, 43, 73
Radiant heat, 171, 174
Radiant heating, from solar hot water, 34
Radiant temperature, 86, 88
Radiation; *See also* Radiative heat exchange
  climate types, 103
  and thermal comfort, 82
Radiation analysis, 34
Radiation impact on surfaces
  absorption factor, 203
  analysis of, 161–162, 163
  HMC Architects Cal Poly Pomona building, 74
  HMC Architects flex school building, 66, 67
  radiative heat exchange, 194–195
  solar geometry, 161–162, 163
  studio projects, advanced, 42, 44, 45, 46
  types of solar protection, 198–200
Radiative heat exchange, 165, 174, 190–209, 210
  air spaces, 185
  building heat balance, 168
  building volume and, 200
  concepts, 190 191
  glazing to surface ratios, 209
  insulating materials, 185
  opaque components, 195–200
    shading devices, 197
    types of solar protection, 198–200
  passive cooling systems, 222, 258–263, 268
    California State University classroom building design, 65
    heat sinks, 222
    principles, 258–260
    UCLA system, 260–263
    Zomeworks Double-play system, 263
  passive heating, heat loss prevention, 269
  prevention of overheating/overcooling, 168, 169
  roof ponds, 247–248
  shading, 144, 197, 209, 210
  solar radiation
    effects on surfaces, 194–195, 196
    factors affecting, 191–194
    sol-air temperature, 196, 197
  surface materials
    absorptive, reflective, emissive, 201–202, 203
    texture and, 201
  surface resistance, 185
  thermal comfort, 84
    climate variables and, 101, 102
    mechanisms of heat transfer, 82, 83
    temperature difference and, 85
  transparent components, 202
  windows and building openings, 207–209
Radiators, 259
Rainfall, 98, 101–102
Rainwater harvesting, 24
  design process roadmap, 21
  studio projects, advanced, 57, 58, 59, 60

Index

Ramirez, Irma, 34
Rating system, window system performance, 29, 34
Recycling, 25, 60, 70, 71
Reduction phase, design process, 17
Reflectance, solar, 67, 71, 191, 192
Reflected light
    analysis of, 162, 163
    light shelves, 50–51
Reflected radiation, 194
    energy flow design, 166, 167
    single-pane windows, 204
    solar gains, 196
Reflective glass, 203, 204
Reflective materials
    barriers and insulators, 185
    radiative heat exchange, 201–202, 203
Reflectivity, colors, 193
Regenerative building, 20
Regional scale, 22
    design process roadmap, 20, 21
    energy simulation tool exercises, 36, 37
Rehbahani, Parinaz, 61, 63, 64
Reiman, Jonathan, 45, 53
Relative humidity
    building bioclimatic chart, 107
    climate analysis
        Climate Consultant wind wheel, 113, 114
        meteorological data, 97, 98
    comfort chart, 225, 226
    heat balance, human body, 85
    passive cooling, 231, 242, 268
    psychrometric charts, 75–78, 79–83; *See also* Psychrometric charts
    and thermal comfort, 82
Renewable energy, 22, 23, 27, 55
Resistance
    conductive heat transfer, 176–177, 180
    time lag, 180
    transmittance (U-value), 178, 179
Resistive insulators, 185
Resurgence, 12, 13
Reuse, 24
    materials, 8, 24
    water; *See* Water reuse
Richeson, Marcus, 36
Ro, Brandon, 61, 62
Rocky Mountain Institute, 14
Roller shades, 145
Roof
    heat exchange
        conductive, 184, 185
        heat gain in cold climates, 171
        longwave radiation losses, 197
        radiant, 260–261, 271
    light sources, 49, 195; *See also* Skylights
    low-energy envelope design exercise, 43
    solar energy conduction, 195
    wind, designing for, 220
Roof monitors, daylighting, 49
Roof ponds, 246–257, 258
    Cal Poly Pomona
        modular roof pond, 255–256, 257, 258
        smart roof pond with floating insulation, 254–255
    in hot humid climates, 253–254
    passive heating, 281–282
    UCLA, 250–253
    University of Nevada, 256–257
Roof systems
    green roofs
        HMC Architects flex school building, 67, 71
        low-energy envelope design exercise, 43
        solar protection, types of, 198
    passive cooling, 268
        green cooling, 236–242
        roof ponds; *See* Roof ponds
    solar radiation
        latitude and, 163, 208–209
        types of solar protection, 198, 199

S

Safeclimate, 12, 13
Safe water, 21, 22, 24
Sagherian, Greg, 32
San Diego County, 35
Sawtooth elements, 49
Scale, 22
    design process roadmap, 20, 21
    energy simulation tool exercises, 36–38
    studio projects, beginning year, 25, 26, 27
Schrotinger, Jillian, 51, 58
Screens, 219
Sealing, heat loss prevention, 269, 270
Seasonality distribution, vegetation, 101
Seasons; *See also* Cold climates
    climate analysis, 102
    daylighting, 32, 52
    low-energy envelope design exercise, 43
    optimal comfort zones for different airspeeds, 225–226
    prevention of overheating/overcooling, 169
    radiation impact on surfaces, 46
    smart ventilation model system, 235–236
    solar and site analysis, 39, 40
    solar energy balance, 205
    solar geometry, 137, 138
        altitude of sun, 139
        solar impacts, 163
    solar heating, 271
    solar radiation summers versus winters, 208
    windows and shading design, 44, 47, 48, 49, 50, 150, 151, 152, 154

Sensible heat
  air movement and, 211, 212, 213
  change in latent heat, 171, 172
  convective heat exchange, 211, 212, 213
  emittance and, 191
  and enthalpy, 77, 78
  humidity and, 242
  passive cooling, 221, 242; *See also* Ventilation
  psychrometric charts; *See* Psychrometric charts
  radiation conversion to, 191
Sensors, low-flow fixtures, 56
Sequester classification of buildings, 19–20
Sequestration, carbon, 1; *See also* Sinks, carbon sequestration
Shades, 146, 186
Shading
  CSUMB classroom building, 65
  HMC Architects designs
    Cal Poly Pomona building, 74
    flex school building, 65, 68, 71
  low-energy envelope design exercise, 43
  neighborhood design, 40–41
  radiative heat exchange, 197–200, 209, 210
    radiation impact on surfaces, 42
    types of solar protection, 198–200
  smart ventilation effects, 235–236, 237
  solar and site analysis, 39, 40, 158, 162
  and solar heat gain coefficient, 204
  studio projects, advanced, 44
Shading design
  solar geometry, 144–146
    horizontal shadow angle (HSA), 146, 147
    shadow angle protractor, 149
    site analysis, 158
    southeast facing window design, 155–157, 158, 159
    south facing window design, 150–154, 155, 156, 157
    vertical shadow angle (VSA), 146–147, 148
  sundials, 157–158, 160
  sun path diagrams, 142
Shading devices, 197–200
Shadow angle protractor, 149, 151
Shadow angles
  horizontal, 146, 147
  vertical, 146–147, 148
Shadow mask
  southeast facing window, 156, 159
  south facing window, 153, 154, 156
Shadows, 97
Shape of building; *See* Geometry
Shelter, 21, 22, 24
Sheltered entries, 63
SHGC (solar heat gain coefficient), 203, 204, 207

Shivering, 95
Shortwave radiation, 194, 204, 206
Showers, 244, 245–246, 247, 253
Shutters, 145, 186
Sidelighting, 49
Sidhu, Sandeesh, 32
Siding, 44
Silica aerogel, 206
SI measurement units, 165
Simulation tool exercises
  studio projects, advanced, 36–38
  studio projects, beginning year, 27
Single-glazed ponds, 282
Single-glazed windows
  design strategy effects on emissions, 286, 287, 288
  passive cooling, 284, 285, 286
  solar radiation, 207, 208
Single-pane windows, 126, 203, 204
Sinks, carbon sequestration, 22
  classification of buildings, 19–20
  design process roadmap, 20, 21
  implementation of, 19
  recycling and, 60
Sinks, heat; *See* Heat sinks; Storage, energy; Thermal mass
SIP (structural insulated panels), 42, 56, 63, 284
Site analysis and design considerations, 22; *See also* Surroundings, buildings and objects
  design process roadmap, 20, 21
  energy simulation tool exercises, 36, 37
  operations and energy emissions reduction, 23
  solar geometry, 158–159
  studio projects, advanced, 37, 39, 40
  studio projects, beginning year, 25, 26
Site-specific climatic conditions, 99–100
Skin, building
  and airflow, 215–219
  energy simulation tool exercises, 36, 37
  heat loss through, 269, 271
  low-energy envelope design exercise, 43
  radiative heat exchange
    materials selection, 201–202, 203
    surface roughness and heat exchange, 201
  solar radiation modeling software, 200, 201
  thermal transfer coefficient, 188
Skin temperature, human body
  heat generation and temperature regulation, 79, 80; *See also* Physiology/heat balance
  mean radiant temperature, 191
  mechanisms of heat flow, 101
Sky
  cloudy/clear; *See* Cloud cover
  longwave radiation, energy flow design, 166, 167

Index

radiative heat exchange, 258, 259
solar geometry, 137, 139, 140, 141; *See also* Solar geometry
upper atmosphere as heat sink, 221, 222, 258
Sky facing surfaces, heat loss via, 269
Skylights
   glass coatings, 205
   orientation of building, 209
   solar gain, 202, 203
   studio projects, advanced, 30, 49, 63
Skytherm System, 249, 256–257, 259–260
Sliding panels, shading devices, 145, 210
Slope, roof/building elements, 118, 261
   designing for wind, 220
   solar gain factor, 202
Slopes
   and incident solar radiation, 137, 138
   precipitation data and, 97
Smart Controls and Thermal Comfort (SCAT), 94
Smart thermostat, 235
Smart ventilation
   experimental conditions, 231–235
   shading effects, 235–236, 237
Smart ventilation system, 214
Smart Window, 34
Social environment, 19
Society of Building Science Educators (SBSE), 29, 34
Software; *See* Digital tools/software
Soil
   heat capacity, 180
   as heat sink, 221, 222, 263–268
Sol-air temperature, 196
Solar analysis
   studio projects, advanced, 37, 39, 40, 41
   windows and shading design, 44, 47–49, 50
Solar design, 22
   design process roadmap, 21
   and emissions, 23, 284–288
   HMC Architects projects
      Cal Poly Pomona building, 65, 74
      CSUMB classroom building, 63, 65, 73
      flex school building, 66, 71
Solar energy/photovoltaic systems, 22
   climate variables, 98
   design process roadmap, 21
   site analysis, 39, 158–159
   solar radiation impact, 161
   studio projects, advanced, 43, 44, 55
   studio projects, beginning year, 28
Solar exposure
   building heat balance, 167–168
   site-specific conditions, 99
Solar gain
   cold climates, 194
   conductive heat transfer, 184

heat balance equations, 166, 167
HMC Architects flex school building, 68
limiting; *See* Shading
material characteristics, 191
passive cooling
   green cooling model, 239, 241
   night ventilation considerations, 230
   smart ventilation model system, 233
radiation impact on surfaces, 42
single-pane windows, 204
solar radiation impact, 161
studio projects, advanced, 53
warm climates, 194–195
Solar gain factor, 202
Solar geometry, 22, 137–163
   design, 146–157, 158, 159
   orientation of building, 162–163
   radiation impact on surfaces, 161–162, 163
   shading, 144–146
   shading design
      horizontal shadow angle (HSA), 146, 147
      shadow angle protractor, 149
      southeast facing window design, 155–157, 158, 159
      south facing window design, 150–154, 155, 156, 157
      vertical shadow angle (VSA), 146–147, 148
   site analysis, 158–159, 161, 162
   solar charts, 140–144
      horizontal sun path diagram, 142, 143, 144
      vertical sun path diagram, 141, 142, 143
   sundials, 157–158, 160
   sun in sky vault, 137, 139, 140, 141
      azimuth and altitude, 139, 141
      declination and hour angle, 137, 139, 140
Solar heat flow; *See also* Solar gain; Solar heating
   building heat balance, 168
   energy flow design, 166, 167
   prevention of overheating/overcooling, 168, 169
Solar heat gain coefficient (SHGC), 203, 204, 207, 232
Solar heating, 270–288
   building bioclimatic charts, 109
   design strategies, effects on emissions, 284–288
   direct gain, 54, 273–275
   form factor, 190
   indirect gain, 275–284
      roof ponds, 281–282
      sun rooms/sunspaces, 282–284
      Trombe and mass walls, 276–279
      water walls, 279–281
   studio projects, advanced, 54

Solar hot water; *See* Water heating
Solarium, 282–284
Solar position
    azimuth and altitude, 139, 141
    Climate Consultant sun-shading chart, 44
Solar protractor, 149, 150, 151, 155–156, 158
Solar radiation
    climate analysis
        heat transfer mechanisms, 101–102
        meteorological data, 97, 98
    conductive heat exchange, 184
    earth coupling, ground/soil as heat sink, 263–267
    energy flow design, 166, 167
    heat balance of buildings, 170–171
    neighborhood design, 40
    passive cooling, 258
        green roofs, 237
        smart ventilation model system, 231–232, 234
    radiative heat exchange
        effects on surfaces, 194–195, 196
        factors affecting, 191–194
    seasonal changes, 137
    and thermal comfort, 102
    windows and shading, 44
Solar reflectance, 67, 71
Solar shadow index, 122–123
Solid-liquid phase change, 171–172
Solstices, 137, 140
Southeast facing window, shading design, 155–157, 158–159
Southern California Edison, 36
Southern California Factor, 15
Southern hemisphere, 137, 140
South facing slopes, 137
South facing window, shading design, 150–154, 155, 156, 157
South pole, 138
Space heating, solar, 54, 222
Space utilization
    and airflow, 215–219
    with solar heating systems, 272
Spain, 99, 121–123
Spanish colonial architecture, 124–125, 126, 127
Specific heat
    air, 167, 170, 212
    energy storage capacity of materials, 187
    heat balance equations, 170
    sensible heat, 171
Spectrally selective coatings, glass, 207, 208
Split flux method, 27
Spraying, roof ponds, 247, 248, 249
Stack ventilation, 43
Standard Oil Company housing, Venezuela, 118
Steady state, thermal, 95, 187
Steady state heat flows, 177

Steel, 180, 203
Stefan-Boltzmann law, 259
Stone, heat capacity, 180
Storage, carbon; *See* Sinks, carbon sequestration
Storage, energy; *See also* Heat sinks; Mass, structural/building; Thermal mass
    cold energy, 261
    latent heat, 171–174
    material characteristics, 187
    passive heating systems
        direct gain, 273–275
        isolated gain systems, 282
        roof ponds, 281–282
        solar systems, 270–271, 272
        Trombe and mass walls, 276–279
        water walls, 278–281
Storage tanks, water, 54, 59
Structural insulated panels (SIP), 42, 56, 63, 284
Structure; *See* Building properties/structure; Mass, structural/building
Structures, adjacent; *See* Massing of buildings; Surroundings, buildings and objects
Studio projects
    advanced, 28–29, 33, 34–60
        climate analysis, 38
        construction, 56
        daylighting, 49–50, 51, 52, 53
        energy simulation tool exercises, 36–38
        envelope, building, 41, 42, 43, 44
        fenestration and shading, 44, 47–49, 50, 51
        geographic distribution of emissions, 38–39, 40
        Green kit, 29
        housing prototypes, 34–35, 36
        irrigation, 56
        light shelves, 50–51
        materials, low-carbon, 55, 56
        modular design, 55
        neighborhood design, 40–41
        plants, use of, 56
        radiation impact on surfaces, 42, 44, 45, 46
        site analysis, 39, 40, 41
        solar analysis, 39, 40, 41
        solar gain, 53
        solar hot water, 53–54
        solar/photovoltaic systems, 55
        ventilation, 51, 53, 54
        waste management, 57, 60
        water conservation, 56–57, 58, 59, 60
    beginning year, 25–28, 29–33
        climate analysis, 28, 29
        daylight analysis, 28, 31, 32, 33
        envelope, building, 28
        photovoltaic design, 28
Sulfur hexafluoride, 1, 2

Index

Summer; *See also* Seasons
    heat transfer mechanisms, 144
    solar geometry, 139
Summer rule, smart ventilation system, 235
Summer solstice, 137, 140
Sun; *See* Solar geometry
Sundials, 157–158, 160
Sun in sky vault, 137, 139, 140, 141
Sun path diagram
    horizontal, 142, 143, 144
    neighborhood design, 40–41
    shading design
        south facing window, 151, 152, 154, 155
        weather data files with, 156
    site analysis, 159, 161, 162
    solar and site analysis, 39, 41
    solar position, 140–144
    vertical, 141, 142, 143
    windows and shading design, 44, 51
Sun rooms/sunspaces, passive heating, 282–284
Sun-shading chart, Climate Consultant, 44
Sunshine duration, meteorological data, 97, 98
Superwindows, 207
Surface area
    conductive heat exchange, 188–189
    form factor, 188–190
    heat balance equations, 170
    solar energy storage medium, 275
    wind, designing for, 220
Surface coefficient, sol-air temperature, 196
Surface resistance
    sol-air temperature, 196
    walls and roofs, 185
Surfaces, outdoor ground, 74
Surfaces of buildings, heat flows; *See also* Envelope, building
    factors affecting solar radiation, 191
    Fourier's equation, 183
    glazing to surface ratios and, 209
    heat loss via, 269
    radiation impact; *See* Radiation impact on surfaces
    radiative heat exchange, 201–202
    and resistance, 177
Surface temperature
    absorptivity and, 197
    green roofs and, 236–237
Surface texture, 185, 193, 201
Surface thermal resistance, walls and roofs, 185
Surroundings, buildings and objects; *See also* Site analysis and design considerations
    and air movement, 215, 219
        design considerations, 220
        and wind patterns, 209
    climate analysis, site-specific conditions, 99–100
    neighborhood design, 40–41
    and passive cooling, 268
    solar and site analysis, 39, 158
        radiation impact on surfaces, 44
        shading by, 158–159, 161
        sun path diagrams, 161
        windows and shading design, 46
    vernacular architecture in temperate climates, 124
    wind patterns, 209
Surrounding surfaces, heat transfer to, 101, 102
Sustainable housing, 34, 35, 36
Sustainable landscaping, 21, 22, 24
Sweat (evaporative cooling), 80, 95, 102, 226

T

Temperate climates, 103
    design strategy effects on emissions, 286
    emissions simulations, 131
    Olgyay building bioclimatic chart and, 108
    operations emissions, 134–136
    varieties of, 123
    vernacular architecture, 124–127, 128
Temperate zone, studio projects, 34–35
Temperature
    body; *See* Physiology/heat balance
    climate analysis
        classification systems, 101
        climate types, 103
        factors affecting design, 99
        heat transfer mechanisms, 101–102
        meteorological data, 97, 98, 101
    cyclic changes; *See* Day-night cycles; Seasons
    global warming, 3–6
    heat capacity and specific heat capacity, 178–179
    measurement units, 165
    operative, 86, 88
    psychrometric charts; *See* Psychrometric charts
    solar radiation impact, 161
    thermal comfort; *See* Thermal comfort
    thermal neutrality, 90, 92, 93–94, 95
    timetable of bioclimatic needs, 112, 113
    variations in; *See* Temperature swings
    wavelength of radiation and, 190
    windows and shading design, 44
        south facing window, 150, 152
        weather data files, 156
Temperature control
    emissions reduction, 23
    operations emissions, 9
Temperature difference, indoor-outdoor
    conductive heat exchange
        cold climates, 195
        decrement factor, 179–180
        delay and reduction of temperature swings, 187–188

convective heat exchange, 210–220; *See also* Convective heat exchange
   closing building, 214
   controlling air exchange, 213–214
   opening building, 214–220
   sealing, 213–214
heat exchange
   heat balance equations, 170–171
   heat balance of building, 167
   thermal mass as insulator, 181
   time lag, 182
passive cooling
   nocturnal ventilation, 230–231
   smart ventilation model system, 233
windows and shading design; *See* Window systems
Temperature difference, surfaces
   conductive heat exchange, 183, 184
   energy storage capacity of materials, 187
Temperature swings
   delay and reduction of, 187–188
   direct gain heating systems, 274
   energy storage capacity of materials, 187
   glazing to surface ratios and, 209
   heat exchange
      decrement factor, 179–180, 183
      heat balance equations, 170–171
      time lag, 182
   indoor, solar heating and, 272
   passive cooling
      green cooling model, 237, 241
      nocturnal ventilation, 228–230
      radiative heat exchange, 260–261
   passive heating
      direct gain systems, 275
      Trombe and mass walls, 278–279
Terrain, 99–100, 220
   climate analysis, 99–100
   ground as heat sink, 263–267
   slope, and incident solar radiation, 137, 138
TerraPass, 12
Texture/roughness
   radiative heat exchange, 193, 201
   surface resistance, 185
Thatch, 116
Thawing, latent heat of fusion, 171–172
Thermal acceptability ratings, 94
Thermal (heat) balance equations, 166–167, 170–171
Thermal break, window systems, 207, 208
Thermal bridges, fins and, 199–200
Thermal capacity; *See* Heat capacity and specific heat capacity
Thermal comfort, 22, 75–96
   ASHRAE definition, 78
   climate and, 97–103

building bioclimatic charts, 106–111
climate variables, 97–99
convective heat transfer, 214–220
cooling systems; *See* Cooling, passive
designing for
   design process roadmap, 21
   HMC Architects Cal Poly Pomona building, 74
   HMC Architects CSUMB classroom building, 65, 73
   HMC Architects flex school building, 67
developed versus undeveloped countries, 109–111
digital climate analysis tools, 112
emissions reduction, 23
environmental and comfort indices, 86, 88–89
evaporative cooling, 246
heat balance, human body, 79–82, 83, 84–86, 87, 88
heating systems; *See* Heating, passive
low-energy envelope design exercise, 41
models of comfort, 89–95
   adaptive, 90, 92–95
   physiological, 89–90, 91
perception of comfort, 95–96
psychrometrics, 75–78, 79–83
shading, 144
vernacular architecture; *See* Vernacular architecture
windows and shading design, 44; *See also* Window systems
Thermal coupling, 239
Thermal equilibrium, 165, 258
Thermal gain, 184
Thermal inertia; *See* Thermal mass
Thermal lag; *See* Time delay/thermal lag
Thermal mass; *See also* Mass, structural/building; Storage, energy
   building bioclimatic charts, 110
   direct gain heating, 54
   heat balance of buildings, 170–171
   heat capacity, 178–179, 181
   passive cooling systems
      green cooling model, 237, 242
      nocturnal ventilation, 229, 230
      smart ventilation model system, 234, 235
      UCLA system, 262–263
   passive heating systems, 269
      direct gain, 273–275
      indirect gain systems, 275–284
      solar systems, 270–271, 272
   phase change materials, 174
   and temperature swing, 120
   Zomeworks Double-play system, 263
Thermal neutrality, 92

Index

Thermal neutrality temperature, 90, 92, 93–94, 95
Thermal performance
    shading design, 156–157
    studio projects, beginning year, 26
    Tijuana low-cost sustainable housing prototype project, 34
Thermal transfer coefficient, 188
Thermal transmittance (U-value); See Transmittance, thermal
Thermistors, 231
Thermometer
    globe, 191
    sensible heat measurement, 171
Thermostat, 235
Thickness, material; See also Geometry
    heat exchange
        and conductance, 175
        conductive, 183
        decrement factor, 183
        and resistance, 176, 177
        time lag, 182
    temperature reduction using high-density and high-thermal capacity materials, 188
    Trombe and mass walls, 277–278
Thickness of architectural elements, conductive heat exchange, 186
Three-dimensional models
    climate effects simulations, 132
    radiation impact on surfaces, 42
    solar radiation on building envelopes, 200, 201
Tijuana low-cost sustainable housing prototype project, 34, 35, 36
    air flow/air movement analysis, 54
    green cooling, 239
Time delay/thermal lag
    conductive heat exchange, 179, 180, 182
    passive heating systems, 278–279
Timetable of bioclimatic needs, 112, 113
Tinted glass, 203, 204, 205
Titanium dioxide paints, 192
Toldo, 121, 122, 198
Tolios, Dimitrios, 61, 63, 64
Tools; See also Digital tools/software
    studio projects, advanced, 36–38
    studio projects, beginning year, 25–26, 28
Toplighting
    daylighting, 49
    and thermal gains, 209
Topography; See Terrain
Towers, 65, 244, 245
Traditional architecture; See Vernacular architecture
Traditional design, 17, 18
Tragish, Emily, 33

Transmittance, thermal (U-value), 85–86, 167, 177–178, 179
    decrement factor, 180
    Fourier's equation, 183
    glazing systems, 206
    green roofs and, 236–237
    heat balance equations, 170
    surface resistance, 185
    windows, 186, 205
Transmittance, visible (VT), 192, 194, 204, 207
Transparency/transparent elements, radiative heat transfer, 191, 192, 194, 195, 202
Transportation emissions
    carbon counting tools, recommended, 15
    climate effects simulations, 134
    studio projects, advanced, 37
Trees
    and airflow, 220
    site analysis, 158
    and solar radiation, 137, 139
Trellis, 198
Triple glazing; See Multilayer glazing
Trombe walls, 276–279
TSA (sol-air temperature), 196
Tucker, Tyler, 43, 47
Turbines, 9, 57

U

Uganda, GreenKit project, 29
U.K. Building Research Establishment (BRE), 27
Ultraviolet radiation, direct gain heating systems, 274
Ultraviolet transmission, 205
Underground systems, ground/soil as heat sink, 264
Undeveloped versus developed countries, comfort zones in, 109–111
United Kingdom, climate data sets, 99
United Nations Environment Programme (UNEP), 2, 9
United States
    climate data sets, 99
    climate zones and energy codes, 103–104, 105
    emissions data sources, 132
    GreenKit project, 29
    natural ventilation strategies in, 230
United States Environmental Protection Agency (EPA) eGRID database, 132
United States Environmental Protection Agency (EPA) Power Profile, 132
University of California, Los Angeles
    evaporative cooling systems
        roof pond, Givoni-LaRoche, 249
        roof pond, Givoni-La Roche, 250–253
        shower, 246, 247
    radiant cooling system, 260–263
    smart ventilation system, 231–236

University of Nevada roof ponds, 256–257
Upper atmosphere, as heat sink, 221, 222, 258
Urban scale
    design process roadmap, 20, 21
    studio projects, advanced, 36
    studio projects, beginning year, 25, 26
U.S. Environmental Protection Agency funding, 28
User controls; *See* Control systems
User interface, design tools, 18
U.S. Green Building Council (USGBC) competition, 35
U-value; *See* Transmittance, thermal

V

van der Linden, P.J., 3, 7
Van Leuween, Garret, 40
Vapor pressure, 75
    building bioclimatic charts, 108
    and nocturnal ventilative cooling, 228
Vasari, 162
Vegetation
    and airflow, 220
    climate analysis, site-specific conditions, 99
    climate classification systems, 100–101
    green cooling model, 236–242
    shading systems, 198, 200
Venetian blinds, 235
Venezuela, 111
    corporate housing, 117–118
    GreenKit project, 29
    informal settlements, radiant cooling, 260–261
    roof ponds, 253–254
    vernacular architecture
        cold climates, 129
        hot dry climates, 120–121
        hot humid climates, 116–118
        temperate climates, 124, 125, 127
Ventilation
    air spaces, 185
    building bioclimatic charts, 109
    climate variables, 98
    convective heat exchange, 210, 211, 213–220
    design strategy effects on emissions, 288
    energy flow design, 166, 167
    form factor, 189
    function and requirements, 223–224
    heat balance equations, 167, 170
    heat transfer mechanisms, 144
    HMC Architects designs
        CSUMB classroom building, 65, 73
        flex school building, 65, 69
    low-energy envelope design exercise, 42, 43
    neighborhood design, 41
    passive cooling; *See* Ventilatory cooling
    shading, 144
    studio projects
        advanced, 33, 34, 37, 51, 53, 54
        beginning year, 25, 26
    vernacular architecture; *See* Vernacular architecture
Ventilation conductance, 167
Ventilatory cooling
    applicability, 267–268
    comfort, 221, 224–226, 227
    green cooling, 236–242
    nocturnal, 221, 222, 226, 227, 228–237, 242, 268
    smart ventilation model system, 231–236, 237
        experimental conditions, 231–235
        shading effects, 235–236, 237
Vents, Trombe walls, 276, 279
Venturi effect, 216
Verandas, 124, 125
Vernacular architecture
    climate, 114–129
        cold climates, 128–129
        hot dry climates, 119–123
        hot humid climates, 115–118, 119
        temperate climates, 124–127, 128
    cold climates, 189–190
Vertical elements
    convective heat exchange, 212
    shadow angle protractor, 149, 151
Vertical shadow angle (VSA)
    shading design, 146–147, 148
        southeast facing window, 156
        south facing window, 151, 153–154
        shadow angle protractor, 149, 150
Vertical sun path diagram, 140
    shading design, south facing window, 150, 152
    site analysis, 159, 162
    solar charts, 141, 142, 143
Vertical windows and light sources, 49
Very simple design (VSD) tools, 18
Vinyl window systems, 207, 208
Visible transmittance (VT), 192, 194, 204, 207
Volume
    building form factor, 188–190
    and energy storage capacity of materials, 187
Volumetric loss, 188
Volumetric mass of air, convective heat exchange, 211
Volumetric specific heat of air, 170
VSA; *See* Vertical shadow angle
VT (visible transmittance), 192, 194, 204, 207

W

Walls
    climate effects on emissions simulations, 131
    convective heat flows, 211, 212

Index

passive cooling, 228
passive heating systems
    glazing to surface ratios and; *See* Glazing systems, glazing to surface ratios
    heat loss, routes of, 269
    heat loss prevention, 271
    indirect gain systems, 275–281
    Trombe and mass walls, 276–279
    water walls, 279–281
solar energy conduction, 195
solar protection, 197, 198–200
surface thermal resistance, 185
thermal balance in cold climates, 171
thermal mass and radiant cooling, UCLA system, 263
Warm climates
    heat balance of buildings, 170–171
    solar radiation effects, 194–195
Warm humid climates; *See* Hot humid climates
Waste, energy generation from, 21, 22, 24
Waste control, 24
    design process roadmap, 21
    waste emissions reduction, 25
Waste emissions, 10, 22
    baseline emissions determination, 19
    carbon counting tools, recommended, 15
    climate effects simulations, 133, 134
    design process, 24–25
    design process roadmap, 20, 21
    HMC Architects flex school building, 72
    studio projects, advanced, 37
    studio projects, beginning year, 27
Waste heat, composting process, 57
Waste management, 22
    HMC Architects flex school building, 71
    studio projects, advanced, 57, 60
Waste to nutrient, 22
Water
    carbon counting tools, 15
    evaporative cooling; *See* Evaporative cooling/heat loss
    global warming effects, 5
    greenhouse gas emissions, 10
    heat capacity, 179, 180
    and heat flow, 184
    as heat sink, 221, 222
        cooling of outdoor spaces, 266
        evaporative cooling; *See also* Evaporative cooling/heat loss
        roof ponds; *See* Roof ponds
        Zomeworks Double-play system, 263
Water bodies
    site-specific conditions, 99
    temperate climates, 123
Water conservation, 22
    design process roadmap, 21
    HMC Architects Cal Poly Pomona building, 74

HMC Architects CSUMB classroom building, 73
HMC Architects flex school building, 66, 72
studio projects, advanced, 56–57, 58, 59, 60
Water Conservation Calculator, 12
Water heating
    climate variables, 98
    HMC Architects flex school building, 65, 68, 71
    low-energy envelope design exercise, 43
    solar radiation impact, 161
    studio projects, advanced, 53–54
Water reuse, 21, 22, 30, 57, 61, 62, 66, 73
Water supply related emissions, 56
    climate effects simulations, 133–134
    design process, 20, 21, 22, 24
    studio projects, advanced, 37
    studio projects, beginning year, 27
Water use
    carbon counting tools, 15
    climate variables, 98
    HMC Architects flex school building, 66
    neighborhood design, 40
Water vapor
    global warning potential, 1
    and heat loss potential of evaporation, 242
    relative humidity, 75, 77–78, 80
Water walls, passive heating, 279–281
Watson, D., 168
Wavelength, radiation, 190, 202
WEA file format, 111
Weather, 97; *See also* Climate analysis; Meteorological/weather data
Weather file formats, 99, 111
Weather Tool (Ecotect), 100, 101
Wells, Malcolm, 29
West facing elevations, solar radiation, 162–163, 208
Wet-bulb temperature
    climate variables, 98, 103
        building bioclimatic charts, 107
        graphing software, 100
    passive cooling systems, 268
        evaporative, 242, 243, 244, 245
        roof ponds, 247, 249, 251, 252
        showers, 253
        towers, 244, 245
    psychrometric charts, 77, 81; *See also* Psychrometric charts
Wetted pads, 243, 244, 245
White paint, 192, 203
White plaster, 193
Whitewash, 192
Whole house fan, 42
Wilderness-based checklist for design and construction, 29–30
Win Air, 53

Wind
  airflow optimization, 214–217
  climate analysis
    heat transfer mechanisms, 101–102
    meteorological data, 98
  designing for, 220
  orientation of building, 209
Window systems (fenestration and shading)
  climate and
    emissions simulations, 131
    heat gain in cold climates, 171
    vernacular architecture; *See* Vernacular architecture
  conductive heat exchange
    multilayer glazing; *See* Multilayer glazing
    thermal resistance, 185–186
  convective heat exchange, 215–219; *See also* Ventilatory cooling
  energy flow design, 166, 167
  HMC Architects designs
    Cal Poly Pomona building, 74
    CSUMB classroom building, 65, 73
    flex school building, 65, 68, 69
  passive cooling, 268
    green cooling model, 239
    smart ventilation model system, 231–232, 234, 235–236, 237
    ventilatory cooling, 227, 228
  passive heating, 269
    control of heat loss, 269
    design strategy effects on emissions, 284–285
    direct gain, 273–275
    heat loss prevention, 271
    solar systems, 272
  radiative heat exchange/solar radiation, 207–209
    glazing to surface ratios, 209
    impact of radiation, 161
    modeling tool, 200
    orientation of building, 207–209
    selection criteria, 202–206, 207
    solar gains, 195, 202
    studio projects, advanced, 54

shading, 144
  southeast facing window, 155–157, 158, 159
  south facing window, 150–154, 155, 156, 157
shadow angle protractor, 149
studio projects, advanced, 29, 33, 34, 47–49, 50, 51, 54
  low-energy envelope design exercise, 44
  radiation impact on surfaces, 42
studio projects, beginning year, 28, 31
and thermal comfort, 93
Wind scoop, shower, 246
Wind shadow, 220
Wind tunnel, 26, 33, 53
Wind turbines, 98
Wind wheel, Climate Consultant, 112, 113, 114
Wind/wind patterns
  climate analysis, site-specific conditions, 99
  sheltered entries, 63
  sol-air temperature, 196
  ventilation, 51, 53, 54
Winter; *See also* Heating, passive; Seasons
  heat loss mechanisms, 144
  solar geometry, 139
Winter rule, smart ventilation model system, 236
Winter solstice, 137, 140
Wood
  absorptivity and emissivity, 203
  heat capacity, 180
  low-energy envelope design exercise, 44
  window systems, 207, 208

Y

Yahoo Green Calculator, 12
Yazell, Erin, 29

Z

ZeroCarbonDesign, 35
Zero emissions, ideal baseline, 19
Zerofootprint: Earthhour, 12, 13
Zero footprint: Unilever Go Blue, 12
Zero waste construction, 38, 56, 61, 63, 65, 70
Zomeworks Double-play system, 263